MW00559535

FREE Test Taking Tips DVD Offer

To help us better serve you, we have developed a Test Taking Tips DVD that we would like to give you for FREE. **This DVD covers world-class test taking tips that you can use to be even more successful when you are taking your test.**

All that we ask is that you email us your feedback about your study guide. Please let us know what you thought about it – whether that is good, bad or indifferent.

To get your **FREE Test Taking Tips DVD**, email freedvd@studyguideteam.com with "FREE DVD" in the subject line and the following information in the body of the email:

> a. The title of your study guide.
>
> b. Your product rating on a scale of 1-5, with 5 being the highest rating.
>
> c. Your feedback about the study guide. What did you think of it?
>
> d. Your full name and shipping address to send your free DVD.

If you have any questions or concerns, please don't hesitate to contact us at freedvd@studyguideteam.com.

Thanks again!

PSAT 10 Prep 2020 and 2021

with Practice Tests [Includes Two PSAT 10 Practice Tests]

TPB Publishing

Copyright © 2020 by TPB Publishing

All rights reserved. No part of this publication may be reproduced, distributed, or transmitted in any form or by any means, including photocopying, recording, or other electronic or mechanical methods, without the prior written permission of the publisher, except in the case of brief quotations embodied in critical reviews and certain other noncommercial uses permitted by copyright law.

Written and edited by TPB Publishing.

TPB Publishing is not associated with or endorsed by any official testing organization. TPB Publishing is a publisher of unofficial educational products. All test and organization names are trademarks of their respective owners. Content in this book is included for utilitarian purposes only and does not constitute an endorsement by TPB Publishing of any particular point of view.

Interested in buying more than 10 copies of our product? Contact us about bulk discounts:
bulkorders@studyguideteam.com

ISBN 13: 9781628458374
ISBN 10: 1628458372

Table of Contents

Quick Overview

As you draw closer to taking your exam, effective preparation becomes more and more important. Thankfully, you have this study guide to help you get ready. Use this guide to help keep your studying on track and refer to it often.

This study guide contains several key sections that will help you be successful on your exam. The guide contains tips for what you should do the night before and the day of the test. Also included are test-taking tips. Knowing the right information is not always enough. Many well-prepared test takers struggle with exams. These tips will help equip you to accurately read, assess, and answer test questions.

A large part of the guide is devoted to showing you what content to expect on the exam and to helping you better understand that content. In this guide are practice test questions so that you can see how well you have grasped the content. Then, answer explanations are provided so that you can understand why you missed certain questions.

Don't try to cram the night before you take your exam. This is not a wise strategy for a few reasons. First, your retention of the information will be low. Your time would be better used by reviewing information you already know rather than trying to learn a lot of new information. Second, you will likely become stressed as you try to gain a large amount of knowledge in a short amount of time. Third, you will be depriving yourself of sleep. So be sure to go to bed at a reasonable time the night before. Being well-rested helps you focus and remain calm.

Be sure to eat a substantial breakfast the morning of the exam. If you are taking the exam in the afternoon, be sure to have a good lunch as well. Being hungry is distracting and can make it difficult to focus. You have hopefully spent lots of time preparing for the exam. Don't let an empty stomach get in the way of success!

When travelling to the testing center, leave earlier than needed. That way, you have a buffer in case you experience any delays. This will help you remain calm and will keep you from missing your appointment time at the testing center.

Be sure to pace yourself during the exam. Don't try to rush through the exam. There is no need to risk performing poorly on the exam just so you can leave the testing center early. Allow yourself to use all of the allotted time if needed.

Remain positive while taking the exam even if you feel like you are performing poorly. Thinking about the content you should have mastered will not help you perform better on the exam.

Once the exam is complete, take some time to relax. Even if you feel that you need to take the exam again, you will be well served by some down time before you begin studying again. It's often easier to convince yourself to study if you know that it will come with a reward!

Test-Taking Strategies

1. Predicting the Answer

When you feel confident in your preparation for a multiple-choice test, try predicting the answer before reading the answer choices. This is especially useful on questions that test objective factual knowledge. By predicting the answer before reading the available choices, you eliminate the possibility that you will be distracted or led astray by an incorrect answer choice. You will feel more confident in your selection if you read the question, predict the answer, and then find your prediction among the answer choices. After using this strategy, be sure to still read all of the answer choices carefully and completely. If you feel unprepared, you should not attempt to predict the answers. This would be a waste of time and an opportunity for your mind to wander in the wrong direction.

2. Reading the Whole Question

Too often, test takers scan a multiple-choice question, recognize a few familiar words, and immediately jump to the answer choices. Test authors are aware of this common impatience, and they will sometimes prey upon it. For instance, a test author might subtly turn the question into a negative, or he or she might redirect the focus of the question right at the end. The only way to avoid falling into these traps is to read the entirety of the question carefully before reading the answer choices.

3. Looking for Wrong Answers

Long and complicated multiple-choice questions can be intimidating. One way to simplify a difficult multiple-choice question is to eliminate all of the answer choices that are clearly wrong. In most sets of answers, there will be at least one selection that can be dismissed right away. If the test is administered on paper, the test taker could draw a line through it to indicate that it may be ignored; otherwise, the test taker will have to perform this operation mentally or on scratch paper. In either case, once the obviously incorrect answers have been eliminated, the remaining choices may be considered. Sometimes identifying the clearly wrong answers will give the test taker some information about the correct answer. For instance, if one of the remaining answer choices is a direct opposite of one of the eliminated answer choices, it may well be the correct answer. The opposite of obviously wrong is obviously right! Of course, this is not always the case. Some answers are obviously incorrect simply because they are irrelevant to the question being asked. Still, identifying and eliminating some incorrect answer choices is a good way to simplify a multiple-choice question.

4. Don't Overanalyze

Anxious test takers often overanalyze questions. When you are nervous, your brain will often run wild, causing you to make associations and discover clues that don't actually exist. If you feel that this may be a problem for you, do whatever you can to slow down during the test. Try taking a deep breath or counting to ten. As you read and consider the question, restrict yourself to the particular words used by the author. Avoid thought tangents about what the author *really* meant, or what he or she was *trying* to say. The only things that matter on a multiple-choice test are the words that are actually in the question. You must avoid reading too much into a multiple-choice question, or supposing that the writer meant something other than what he or she wrote.

5. No Need for Panic

It is wise to learn as many strategies as possible before taking a multiple-choice test, but it is likely that you will come across a few questions for which you simply don't know the answer. In this situation, avoid panicking. Because most multiple-choice tests include dozens of questions, the relative value of a single wrong answer is small. As much as possible, you should compartmentalize each question on a multiple-choice test. In other words, you should not allow your feelings about one question to affect your success on the others. When you find a question that you either don't understand or don't know how to answer, just take a deep breath and do your best. Read the entire question slowly and carefully. Try rephrasing the question a couple of different ways. Then, read all of the answer choices carefully. After eliminating obviously wrong answers, make a selection and move on to the next question.

6. Confusing Answer Choices

When working on a difficult multiple-choice question, there may be a tendency to focus on the answer choices that are the easiest to understand. Many people, whether consciously or not, gravitate to the answer choices that require the least concentration, knowledge, and memory. This is a mistake. When you come across an answer choice that is confusing, you should give it extra attention. A question might be confusing because you do not know the subject matter to which it refers. If this is the case, don't eliminate the answer before you have affirmatively settled on another. When you come across an answer choice of this type, set it aside as you look at the remaining choices. If you can confidently assert that one of the other choices is correct, you can leave the confusing answer aside. Otherwise, you will need to take a moment to try to better understand the confusing answer choice. Rephrasing is one way to tease out the sense of a confusing answer choice.

7. Your First Instinct

Many people struggle with multiple-choice tests because they overthink the questions. If you have studied sufficiently for the test, you should be prepared to trust your first instinct once you have carefully and completely read the question and all of the answer choices. There is a great deal of research suggesting that the mind can come to the correct conclusion very quickly once it has obtained all of the relevant information. At times, it may seem to you as if your intuition is working faster even than your reasoning mind. This may in fact be true. The knowledge you obtain while studying may be retrieved from your subconscious before you have a chance to work out the associations that support it. Verify your instinct by working out the reasons that it should be trusted.

8. Key Words

Many test takers struggle with multiple-choice questions because they have poor reading comprehension skills. Quickly reading and understanding a multiple-choice question requires a mixture of skill and experience. To help with this, try jotting down a few key words and phrases on a piece of scrap paper. Doing this concentrates the process of reading and forces the mind to weigh the relative importance of the question's parts. In selecting words and phrases to write down, the test taker thinks about the question more deeply and carefully. This is especially true for multiple-choice questions that are preceded by a long prompt.

9. Subtle Negatives

One of the oldest tricks in the multiple-choice test writer's book is to subtly reverse the meaning of a question with a word like *not* or *except*. If you are not paying attention to each word in the question, you can easily be led astray by this trick. For instance, a common question format is, "Which of the following is...?" Obviously, if the question instead is, "Which of the following is not...?," then the answer will be quite different. Even worse, the test makers are aware of the potential for this mistake and will include one answer choice that would be correct if the question were not negated or reversed. A test taker who misses the reversal will find what he or she believes to be a correct answer and will be so confident that he or she will fail to reread the question and discover the original error. The only way to avoid this is to practice a wide variety of multiple-choice questions and to pay close attention to each and every word.

10. Reading Every Answer Choice

It may seem obvious, but you should always read every one of the answer choices! Too many test takers fall into the habit of scanning the question and assuming that they understand the question because they recognize a few key words. From there, they pick the first answer choice that answers the question they believe they have read. Test takers who read all of the answer choices might discover that one of the latter answer choices is actually *more* correct. Moreover, reading all of the answer choices can remind you of facts related to the question that can help you arrive at the correct answer. Sometimes, a misstatement or incorrect detail in one of the latter answer choices will trigger your memory of the subject and will enable you to find the right answer. Failing to read all of the answer choices is like not reading all of the items on a restaurant menu: you might miss out on the perfect choice.

11. Spot the Hedges

One of the keys to success on multiple-choice tests is paying close attention to every word. This is never truer than with words like almost, most, some, and sometimes. These words are called "hedges" because they indicate that a statement is not totally true or not true in every place and time. An absolute statement will contain no hedges, but in many subjects, the answers are not always straightforward or absolute. There are always exceptions to the rules in these subjects. For this reason, you should favor those multiple-choice questions that contain hedging language. The presence of qualifying words indicates that the author is taking special care with his or her words, which is certainly important when composing the right answer. After all, there are many ways to be wrong, but there is only one way to be right! For this reason, it is wise to avoid answers that are absolute when taking a multiple-choice test. An absolute answer is one that says things are either all one way or all another. They often include words like *every*, *always*, *best*, and *never*. If you are taking a multiple-choice test in a subject that doesn't lend itself to absolute answers, be on your guard if you see any of these words.

12. Long Answers

In many subject areas, the answers are not simple. As already mentioned, the right answer often requires hedges. Another common feature of the answers to a complex or subjective question are qualifying clauses, which are groups of words that subtly modify the meaning of the sentence. If the question or answer choice describes a rule to which there are exceptions or the subject matter is complicated, ambiguous, or confusing, the correct answer will require many words in order to be expressed clearly and accurately. In essence, you should not be deterred by answer choices that seem excessively long. Oftentimes, the author of the text will not be able to write the correct answer without

offering some qualifications and modifications. Your job is to read the answer choices thoroughly and completely and to select the one that most accurately and precisely answers the question.

13. Restating to Understand

Sometimes, a question on a multiple-choice test is difficult not because of what it asks but because of how it is written. If this is the case, restate the question or answer choice in different words. This process serves a couple of important purposes. First, it forces you to concentrate on the core of the question. In order to rephrase the question accurately, you have to understand it well. Rephrasing the question will concentrate your mind on the key words and ideas. Second, it will present the information to your mind in a fresh way. This process may trigger your memory and render some useful scrap of information picked up while studying.

14. True Statements

Sometimes an answer choice will be true in itself, but it does not answer the question. This is one of the main reasons why it is essential to read the question carefully and completely before proceeding to the answer choices. Too often, test takers skip ahead to the answer choices and look for true statements. Having found one of these, they are content to select it without reference to the question above. Obviously, this provides an easy way for test makers to play tricks. The savvy test taker will always read the entire question before turning to the answer choices. Then, having settled on a correct answer choice, he or she will refer to the original question and ensure that the selected answer is relevant. The mistake of choosing a correct-but-irrelevant answer choice is especially common on questions related to specific pieces of objective knowledge. A prepared test taker will have a wealth of factual knowledge at his or her disposal, and should not be careless in its application.

15. No Patterns

One of the more dangerous ideas that circulates about multiple-choice tests is that the correct answers tend to fall into patterns. These erroneous ideas range from a belief that B and C are the most common right answers, to the idea that an unprepared test-taker should answer "A-B-A-C-A-D-A-B-A." It cannot be emphasized enough that pattern-seeking of this type is exactly the WRONG way to approach a multiple-choice test. To begin with, it is highly unlikely that the test maker will plot the correct answers according to some predetermined pattern. The questions are scrambled and delivered in a random order. Furthermore, even if the test maker was following a pattern in the assignation of correct answers, there is no reason why the test taker would know which pattern he or she was using. Any attempt to discern a pattern in the answer choices is a waste of time and a distraction from the real work of taking the test. A test taker would be much better served by extra preparation before the test than by reliance on a pattern in the answers.

5

FREE DVD OFFER

Don't forget that doing well on your exam includes both understanding the test content and understanding how to use what you know to do well on the test. We offer a completely FREE Test Taking Tips DVD that covers world class test taking tips that you can use to be even more successful when you are taking your test.

All that we ask is that you email us your feedback about your study guide. To get your **FREE Test Taking Tips DVD**, email freedvd@studyguideteam.com with "FREE DVD" in the subject line and the following information in the body of the email:

- The title of your study guide.
- Your product rating on a scale of 1-5, with 5 being the highest rating.
- Your feedback about the study guide. What did you think of it?
- Your full name and shipping address to send your free DVD.

Introduction to the PSAT 10

Function of the Test

The Preliminary SAT 10 (PSAT 10) is an introductory version of the SAT exam. Given by the College Board with support from the National Merit Scholarship Corporation (NMSC), the PSAT is designed to help U.S. students get ready for the SAT or ACT. It also serves as a qualifying measure to identify students for college scholarships, but unlike the PSAT/NMSQT, it is not part of the National Merit Scholarship Program. This is a contest that recognizes and awards scholars based on academic performance. About 50,000 pupils are acknowledged for extraordinary PSAT scores every year. Approximately 16,000 of these students become National Merit Semifinalists, and about half of this group is awarded scholarships.

Over 3.5 million high school students take the PSAT or PSAT 10 every year. Most are high school sophomores or juniors residing in the U.S. However, younger students may also register to take the PSAT. Students who are not U.S. citizens or residents can take the PSAT as well, by locating and contacting a local school that offers it.

Test Administration

The PSAT 10 is offered on various dates in the spring at schools throughout the United States. Some schools will pay all or part of the exam registration fee for their pupils. Since the financial responsibility of the student for the exam is different for each school, it is best to consult the school's guidance department for specifics.

Students who would like multiple chances to take the PSAT can take the PSAT/NMSQT in the fall and the PSAT 10 in the spring.

Students with documented disabilities can contact the College Board to make alternative arrangements to take the PSAT. All reasonable applications are reviewed.

Test Format

The PSAT 10 gauges a student's proficiency in three areas: Reading, Mathematics, and Writing and Language. All the tests that fall under the SAT umbrella (including the PSAT) were redesigned in 2015. The revised PSAT 10 is very similar to the new SAT in substance, structure, and scoring methodology, except that the PSAT does not include an essay. 1520 is the highest possible score for the PSAT.

The reading portion of the PSAT 10 measures comprehension, requiring candidates to read multi-paragraph fiction and non-fiction segments including informational visuals, such as charts, tables and graphs, and answer questions based on this content. Three critical sectors are tested for the math section: Solving problems and analyzing data, Algebra, and complex equations and operations. The writing and language portion requires students to evaluate and edit writing and graphics to obtain an answer that correctly conveys the information given in the passage.

The PSAT 10 contains 139 multiple-choice questions, with each section comprising over 40 questions. A different length of time is given for each section, for a total of two hours, 45 minutes.

Section	Time (In Minutes)	Number of Questions
Reading	60	47
Writing and Language	35	44
Mathematics	70	48
Total	**165**	**139**

Scoring

Scores for the newly revised PSAT are based on a scale of 320 to 1520. Scores range from 160-760 for the Math section and 160-760 for the Reading and Writing and Language combined. The PSAT 10 also no longer penalizes for incorrect answers, as it did in the previous version. Therefore, a student's raw score is the number of correctly answered questions. Score reports also list sub-scores for math, reading and writing on a scale from 8 to 38, in order to give candidates an idea of strengths and weaknesses. Mean, or average, scores received by characteristic U.S. test-takers, are broken down by grade level.

The report ranks scores based on a percentile between 1 and 99 so students can see how they measured up to other test takers. Average (50th percentile), scores range from about 470 to 480 in each section, for a total of 940 to 960. Good scores are typically defined as higher than 50 percent. Scores of 95 percent or higher are in contention for National Merit Semifinalist and Finalist slots, but scholarships usually only go to the top one percent of 10th graders taking the PSAT.

Recent/Future Developments

A redesigned version of the PSAT 10 was launched in October 2015. Changes include a longer total length (2 hours, 45 minutes, versus the previous time of 2 hours, 10 minutes), a total of four multiple-choice answers per question instead of five as in the past, and no guessing penalty, so students earn points based only on questions answered correctly. The revised PSAT 10 also has more of a well-rounded emphasis on life skills and the thinking needed at a college level, incorporating concepts learned in science, history and social studies into the reading, math and writing sections. It is important to note that since the redesigned PSAT is different than in the past, scores on previous tests should not be compared to those taken in the current year.

When the PSAT 10 was revamped, a number of new Services for Students with Disabilities (SSD) regulations occurred as well. For example, the PSAT/NMSQT printed test manual for nonstandard testers (often referred to as the "pink book") is no longer used. Instead, every candidate will use the standard exam booklet unless an alternative design (such as large print, Braille, MP3 Audio, and Assistive Technology Compatible) is requested.

There is also a new option allowing students to save time by completing classifying data prior to the exam by choosing the pre-administration option on the PSAT registration website. And starting in January 2015, the College Board forged new collaborations with five scholarship providers to expand scholarship opportunities earlier in students' high school careers.

Reading Test

The purpose of this guide is to help test takers understand the basic principles of reading comprehension questions contained in the Preliminary SAT/National Merit Qualifying Test (PSAT/NMSQT). Studying this guide will help determine the types of questions that the test contains and how best to address them, provided the test's parameters. This guide is not all-inclusive, and does not contain actual test material. This guide is, and should be used, only as preparation to improve student's reading skills for the PSAT Reading Comprehension section.

Each section addresses key skills test takers need to master in order to successfully complete the Reading portion of the PSAT. Each section is further broken down into sub-skills. All of the topics and related subtopics address testable material. Careful use of this guide should fully prepare test takers for a successful test experience.

Command of Evidence

Command of evidence, or the ability to use contextual clues, factual statements, and corroborative phrases to support an author's message or intent, is an important part of the PSAT/NMSQT. A test taker's ability to parse out factual information and draw conclusions based on evidence is important to critical reading comprehension. The test will ask students to read text passages, and then answer questions based on information contained in them. These types of questions may ask test takers to identify stated facts. They may also require test takers to draw logical conclusions, identify data based on graphs, make inferences, and to generally display analytical thinking skills.

Finding Evidence in a Passage

The basic tenet of reading comprehension is the ability to read and understand text. One way to understand text is to look for information that supports the author's main idea, topic, or position statement. This information may be factual, or it may be based on the author's opinion. This section will focus on the test taker's ability to identify factual information, as opposed to opinionated bias. The PSAT/NMSQT will ask test takers to read passages containing factual information, and then logically relate those passages by drawing conclusions based on evidence.

In order to identify factual information within one or more text passages, begin by looking for statements of fact. Factual statements can be either true or false. Identifying factual statements as opposed to opinion statements is important in demonstrating full command of evidence in reading. For example, the statement *The temperature outside was unbearably hot* may seem like a fact; however, it's not. While anyone can point to a temperature gauge as factual evidence, the statement itself reflects only an opinion. Some people may find the temperature unbearably hot. Others may find it comfortably warm. Thus, the sentence, *The temperature outside was unbearably hot,* reflects the opinion of the author who found it unbearable. If the text passage followed up the sentence with atmospheric conditions indicating heat indices above 140 degrees Fahrenheit, then the reader knows there is factual information that supports the author's assertion of *unbearably hot*.

In looking for information that can be proven or disproven, it's helpful to scan for dates, numbers, timelines, equations, statistics, and other similar data within any given text passage. These types of indicators will point to proven particulars. For example, the statement, *The temperature outside was unbearably hot on that summer day, July 10, 1913,* most likely indicates factual information, even if the reader is unaware that this is the hottest day on record in the United States. Be careful when reading biased words from an author. Biased words indicate opinion, as opposed to fact. See the list of biased words below and keep in mind that it's not an inclusive list:

- Good/bad
- Great/greatest
- Better/best/worst
- Amazing
- Terrible/bad/awful
- Beautiful/handsome/ugly
- More/most
- Exciting/dull/boring
- Favorite
- Very
- Probably/should/seem/possibly

Remember, most of what is written is actually opinion or carefully worded information that seems like fact when it isn't. To say, *duplicating DNA results is not cost-effective* sounds like it could be a scientific fact, but it isn't. Factual information can be verified through independent sources.

The simplest type of test question may provide a text passage, then ask the test taker to distinguish the correct factual supporting statement that best answers the corresponding question on the test. However, be aware that most questions may ask the test taker to read more than one text passage and identify which answer best supports an author's topic. While the ability to identify factual information is critical, these types of questions require the test taker to identify chunks of details, and then relate them to one another.

Displaying Analytical Thinking Skills

Analytical thinking involves being able to break down visual information into manageable portions in order to solve complex problems or process difficult concepts. This skill encompasses all aspects of command of evidence in reading comprehension.

A reader can approach analytical thinking in a series of steps. First, when approaching visual material, a reader should identify an author's thought process. Is the line of reasoning clear from the presented passage, or does it require inference and coming to a conclusion independent of the author? Next, a reader should evaluate the author's line of reasoning to determine if the logic is sound. Look for evidentiary clues and cited sources. Do these hold up under the author's argument? Third, look for bias. Bias includes generalized, emotional statements that will not hold up under scrutiny, as they are not based on fact. From there, a reader should ask if the presented evidence is trustworthy. Are the facts cited from reliable sources? Are they current? Is there any new factual information that has come to light since the passage was written that renders the argument useless? Next, a reader should carefully think about information that opposes the author's view. Do the author's arguments guide the reader to identical thoughts, or is there room for sound arguments? Finally, a reader should always be able to identify an author's conclusion and be able to weigh its effectiveness.

The ability to display analytical thinking skills while reading is key in any standardized testing situation. Test takers should be able to critically evaluate the information provided, and then answer questions related to content by using the steps above.

Making Inferences

Simply put, an inference is making an educated guess drawn from evidence, logic, and reasoning. The key to making inferences is identifying clues within a passage, and then using common sense to arrive at a reasonable conclusion. Consider it "reading between the lines."

One way to make an inference is to look for main topics. When doing so, pay particular attention to any titles, headlines, or opening statements made by the author. Topic sentences or repetitive ideas can be clues in gleaning inferred ideas. For example, if a passage contains the phrase *While some consider DNA testing to be infallible, it is an inherently flawed technique,* the test taker can infer the rest of the passage will contain information that points to problems with DNA testing.

The test taker may be asked to make an inference based on prior knowledge, but may also be asked to make predictions based on new ideas. For example, the test taker may have no prior knowledge of DNA other than its genetic property to replicate. However, if the reader is given passages on the flaws of DNA testing with enough factual evidence, the test taker may arrive at the inferred conclusion that the author does not support the infallibility of DNA testing in all identification cases.

When making inferences, it is important to remember that the critical thinking process involved must be fluid and open to change. While a reader may infer an idea from a main topic, general statement, or other clues, they must be open to receiving new information within a particular passage. New ideas presented by an author may require the test taker to alter an inference. Similarly, when asked questions that require making an inference, it's important to read the entire test passage and all of the answer options. Often, a test taker will need to refine a general inference based on new ideas that may be presented within the test itself.

Author's Use of Evidence to Support Claims

Authors utilize a wide range of techniques to tell a story or communicate information. Readers should be familiar with the most common of these techniques. Techniques of writing are also commonly known as rhetorical devices, and they are some of the evidence that authors use to support claims.

In non-fiction writing, authors employ argumentative techniques to present their opinion to readers in the most convincing way. First of all, persuasive writing usually includes at least one type of appeal: an appeal to logic (logos), emotion (pathos), or credibility and trustworthiness (ethos). When a writer appeals to logic, they are asking readers to agree with them based on research, evidence, and an established line of reasoning. An author's argument might also appeal to readers' emotions, perhaps by including personal stories and anecdotes (a short narrative of a specific event). A final type of appeal, appeal to authority, asks the reader to agree with the author's argument on the basis of their expertise or credentials. Consider three different approaches to arguing the same opinion:

Logic (Logos)

Below is an example of an appeal to logic. The author uses evidence to disprove the logic of the school's rule (the rule was supposed to reduce discipline problems; the number of problems has not been reduced; therefore, the rule is not working) and call for its repeal.

> Our school should abolish its current ban on cell phone use on campus. This rule was adopted last year as an attempt to reduce class disruptions and help students focus more on their lessons. However, since the rule was enacted, there has been no change in the number of disciplinary problems in class. Therefore, the rule is ineffective and should be done away with.

Emotion (Pathos)

An author's argument might also appeal to readers' emotions, perhaps by including personal stories and anecdotes.

The next example presents an appeal to emotion. By sharing the personal anecdote of one student and speaking about emotional topics like family relationships, the author invokes the reader's empathy in asking them to reconsider the school rule.

> Our school should abolish its current ban on cell phone use on campus. If they aren't able to use their phones during the school day, many students feel isolated from their loved ones. For example, last semester, one student's grandmother had a heart attack in the morning. However, because he couldn't use his cell phone, the student didn't know about his grandmother's accident until the end of the day—when she had already passed away, and it was too late to say goodbye. By preventing students from contacting their friends and family, our school is placing undue stress and anxiety on students.

Credibility (Ethos)

Finally, an appeal to authority includes a statement from a relevant expert. In this case, the author uses a doctor in the field of education to support the argument. All three examples begin from the same opinion—the school's phone ban needs to change—but rely on different argumentative styles to persuade the reader.

> Our school should abolish its current ban on cell phone use on campus. According to Dr. Bartholomew Everett, a leading educational expert, "Research studies show that cell phone usage has no real impact on student attentiveness. Rather, phones provide a valuable technological resource for learning. Schools need to learn how to integrate this new technology into their curriculum." Rather than banning phones altogether, our school should follow the advice of experts and allow students to use phones as part of their learning.

Informational Graphics

A test taker's ability to draw conclusions from an informational graphic is a sub-skill in displaying one's command of reading evidence. Drawing conclusions requires the reader to consider all information provided in the passage, then to use logic to piece it together to form a reasonably correct resolution. In this case, a test taker must look for facts as well as opinionated statements. Both should be considered in order to arrive at a conclusion. These types of questions test one's ability to conduct logical and analytical thinking.

Identifying data-driven evidence in informational graphics is very similar to analyzing factual information. However, it often involves the use of graphics in order to do so. In these types of questions, the test taker will be presented with a graph, or organizational tool, and asked questions regarding the information it contains. On the following page, review the pie chart organizing percentages of primary occupations of public transportation passengers in US cities.

This figure depicts the jobs of passengers taking public transportation in U.S. cities. A corresponding PSAT question may have the test taker study the chart, then answer a question regarding the values. For example, is the number of students relying on public transportation greater or less than the number of the unemployed? Similarly, the test may ask if people employed outside the home are less likely to use public transportation than homemakers. Note that the phrase *less likely* may weigh into the reader's

choice of optional answers and that the test taker should look for additional passage data to arrive at a conclusion one way or another.

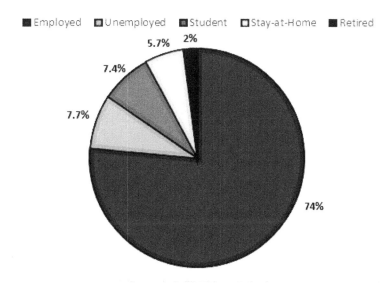

The PSAT/NMSQT will also test the ability to draw a conclusion by presenting the test taker with more than one passage, and then ask questions that require the reader to compare the passages in order to arrive at a logical conclusion. For example, a text passage may describe the flaws in DNA testing, and then describe the near infallibility of it in the next. The test taker then may be required to glean the evidence in both passages, then answer a question such as *the central idea in the first paragraph is . . .* followed by *which choice regarding the infallibility of DNA testing best refutes the previous question?* In this example, the test taker must carefully find a central concept of the flaws of DNA testing based on the two passages, and then rely on that choice to best answer the subsequent question regarding its infallibility.

Words in Context

In order to successfully complete the reading comprehension section of the PSAT/NMSQT, the test taker should be able to identify words in context. This involves a set of skills that requires the test taker to answer questions about unfamiliar words within a particular text passage. Additionally, the test taker may be asked to answer critical thinking questions based on unfamiliar word meaning. Identifying words in context is very much like solving a puzzle. By using a variety of techniques, a test taker should be able to correctly identify unfamiliar words and concepts with ease.

Using Context Clues

A context clue is a hint that an author provides to the reader in order to help define difficult or unique words. When reading a passage, a test taker should take note of any unfamiliar words, and then examine the sentence around them to look for clues to the word meanings.

Let's look at an example:

> He faced a *conundrum* in making this decision. He felt as if he had come to a crossroads. This was truly a puzzle, and what he did next would determine the course of his future.

The word *conundrum* may be unfamiliar to the reader. By looking at context clues, the reader should be able to determine its meaning. In this passage, context clues include the idea of making a decision and of being unsure. Furthermore, the author restates the definition of conundrum in using the word *puzzle* as a synonym. Therefore, the reader should be able to determine that the definition of the word *conundrum* is a difficult puzzle.

Similarly, a reader can determine difficult vocabulary by identifying antonyms. Let's look at an example:

> Her *gregarious* nature was completely opposite of her twin's, who was shy, retiring, and socially nervous.

The word *gregarious* may be unfamiliar. However, by looking at the surrounding context clues, the reader can determine that *gregarious* does not mean shy. The twins' personalities are being contrasted. Therefore, *gregarious* must mean sociable, or something similar to it.

At times, an author will provide contextual clues through a cause and effect relationship. Look at the next sentence as an example:

> The athletes were excited with *elation* when they won the tournament; unfortunately, their off-court antics caused them to forfeit the win.

The word *elation* may be unfamiliar to the reader. However, the author defines the word by presenting a cause and effect relationship. The athletes were so elated at the win that their behavior went overboard and they had to forfeit. In this instance, *elated* must mean something akin to overjoyed, happy, and overexcited.

Cause and effect is one technique authors use to demonstrate relationships. A cause is why something happens. The effect is what happens as a result. For example, a reader may encounter text such as *Because he was unable to sleep, he was often restless and irritable during the day.* The cause is insomnia due to lack of sleep. The effect is being restless and irritable. When reading for a cause and effect relationship, look for words such as "if", "then", "such", and "because." By using cause and effect, an author can describe direct relationships, and convey an overall theme, particularly when taking a stance on their topic.

An author can also provide contextual clues through comparison and contrast. Let's look at an example:

> Her *torpid* state caused her parents, and her physician, to worry about her seemingly sluggish well-being.

The word *torpid* is probably unfamiliar to the reader. However, the author has compared *torpid* to a state of being and, moreover, one that's worrisome. Therefore, the reader should be able to determine that *torpid* is not a positive, healthy state of being. In fact, through the use of comparison, it means sluggish. Similarly, an author may contrast an unfamiliar word with an idea. In the sentence *Her torpid state was completely opposite of her usual, bubbly self,* the meaning of *torpid*, or sluggish, is contrasted with the words *bubbly self.*

A test taker should be able to critically assess and determine unfamiliar word meanings through the use of an author's context clues in order to fully comprehend difficult text passages.

Relating Unfamiliar Words to Familiar Words

The PSAT/NMSQT will test a reader's ability to use context clues, and then relate unfamiliar words to more familiar ones. Using the word *torpid* as an example, the test may ask the test taker to relate the meaning of the word to a list of vocabulary options and choose the more familiar word as closest in meaning. In this case, the test may say something like the following:

Which of the following words means the same as the word *torpid* in the above passage?

Then they will provide the test taker with a list of familiar options such as happy, disgruntled, sluggish, and animated. By using context clues, the reader has already determined the meaning of *torpid* as slow or sluggish, so the reader should be able to correctly identify the word *sluggish* as the correct answer.

One effective way to relate unfamiliar word meanings to more familiar ones is to substitute the provided word answer options with the unfamiliar word in question. Although this will not always lead to a correct answer every time, this strategy will help the test taker narrow answer options. Be careful when utilizing this strategy. Pay close attention to the meaning of sentences and answer choices because it's easy to mistake answer choices as correct when they are easily substituted, especially when they are the same part of speech. Does the sentence mean the same thing with the substituted word option in place or does it change entirely? Does the substituted word make sense? Does it possibly mean the same as the unfamiliar word in question?

How an Author's Word Choice Shapes Meaning, Style, and Tone

Authors choose their words carefully in order to artfully depict meaning, style, and tone, which is most commonly inferred through the use of adjectives and verbs. The *tone* is the predominant emotion present in the text, and represents the attitude or feelings that an author has towards a character or event.

To review, an adjective is a word used to describe something, and usually precedes the noun, a person, place, or object. A verb is a word describing an action. For example, the sentence "The scary woodpecker ate the spider" includes the adjective "scary," the noun "woodpecker," and the verb "ate." Reading this sentence may rouse some negative feelings, as the word "scary" carries a negative charge. The *charge* is the emotional connotation that can be derived from the adjectives and verbs and is either positive or negative. Recognizing the charge of a particular sentence or passage is an effective way to understand the meaning and tone the author is trying to convey.

Many authors have conflicting charges within the same text, but a definitive tone can be inferred by understanding the meaning of the charges relative to each other. It's important to recognize key conjunctions, or words that link sentences or clauses together. There are several types and subtypes of conjunctions. Three are most important for reading comprehension:

- *Cumulative conjunctions* add one statement to another.
- Examples: and, both, also, as well as, not only
- e.g. The juice is sweet *and* sour.
- *Adversative conjunctions* are used to contrast two clauses.
- Examples: but, while, still, yet, nevertheless

- e.g. She was tired, *but* she was happy.
- *Alternative conjunctions* express two alternatives.
- Examples: or, either, neither, nor, else, otherwise
- e.g. He must eat, *or* he will die.

Identifying the meaning and tone of a text can be accomplished with the following techniques:

- Identify the adjectives and verbs.
- Recognize any important conjunctions.
- Label the adjectives and verbs as positive or negative.
- Understand what the charge means about the text.

To demonstrate these steps, examine the following passage from the classic children's poem, "The Sheep":

> Lazy sheep, pray tell me why
>
> In the pleasant fields you lie,
>
> Eating grass, and daisies white,
>
> From the morning till the night?
>
> Everything can something do,
>
> But what kind of use are you?
>
> –Taylor, Jane and Ann. "The Sheep."

This selection is a good example of conflicting charges that work together to express an overall tone. Following the first two steps, identify the adjectives, verbs, and conjunctions within the passage. For this example, the adjectives are <u>underlined</u>, the verbs are in **bold**, and the conjunctions *italicized*:

> <u>Lazy</u> sheep, pray **tell** me why
>
> In the <u>pleasant</u> fields you **lie**,
>
> **Eating** grass, and daisies <u>white,</u>
>
> From the morning till the night?
>
> Everything can something do,
>
> *But* what kind of use are you?

For step three, read the passage and judge whether feelings of positivity or negativity arose. Then assign a charge to each of the words that were outlined. This can be done in a table format, or simply by writing a + or − next to the word.

The word <u>lazy</u> carries a negative connotation; it usually denotes somebody unwilling to work. To **tell** someone something has an exclusively neutral connotation, as it depends on what's being told, which has not yet been revealed at this point, so a charge can be assigned later. The word <u>pleasant</u> is an

inherently positive word. To **lie** could be positive or negative depending on the context, but as the subject (the sheep) is lying in a pleasant field, then this is a positive experience. **Eating** is also generally positive.

After labeling the charges for each word, it might be inferred that the tone of this poem is happy and maybe even admiring or innocuously envious. However, notice the adversative conjunction, "but" and what follows. The author has listed all the pleasant things this sheep gets to do all day, but the tone changes when the author asks, "What kind of use are you?" Asking someone to prove their value is a rather hurtful thing to do, as it implies that the person asking the question doesn't believe the subject has any value, so this could be listed under negative charges. Referring back to the verb **tell**, after reading the whole passage, it can be deduced that the author is asking the sheep to tell what use the sheep is, so this has a negative charge.

+	−
• Pleasant • Lie in fields • From morning to night	• Lazy • Tell me • What kind of use are you

Upon examining the charges, it might seem like there's an even amount of positive and negative emotion in this selection, and that's where the conjunction "but" becomes crucial to identifying the tone. The conjunction "but" indicates there's a contrasting view to the pleasantness of the sheep's daily life, and this view is that the sheep is lazy and useless, which is also indicated by the first line, "lazy sheep, pray tell me why."

It might be helpful to look at questions pertaining to tone. For this selection, consider the following question:

The author of the poem regards the sheep with a feeling of what?
  a. Respect
  b. Disgust
  c. Apprehension
  d. Intrigue

Considering the author views the sheep as lazy with nothing to offer, Choice A appears to reflect the opposite of what the author is feeling.

Choice B seems to mirror the author's feelings towards the sheep, as laziness is considered a disreputable trait, and people (or personified animals, in this case) with unfavorable traits might be viewed with disgust.

Choice C doesn't make sense within context, as laziness isn't usually feared.

Choice D is tricky, as it may be tempting to argue that the author is intrigued with the sheep because they ask, "pray tell me why." This is another out-of-scope answer choice as it doesn't *quite* describe the feelings the author experiences and there's also a much better fit in Choice B.

Rhetoric and Synthesis

Rhetoric

The PSAT/NMSQT will test a reader's ability to identify an author's use of rhetoric within text passages. Rhetoric is the use of positional or persuasive language to convey one or more central ideas. The idea behind the use of rhetoric is to convince the reader of something. Its use is meant to persuade or motivate the reader. An author may choose to appeal to their audience through logic, emotion, the use of ideology, or by conveying that the central idea is timely, and thus, important to the reader. There are a variety of rhetorical techniques an author can use to achieve this goal.

An author may choose to use traditional elements of style to persuade the reader. They may also use a story's setting, mood, characters, or a central conflict to build emotion in the reader. Similarly, an author may choose to use specific techniques such as alliteration, irony, metaphor, simile, hyperbole, allegory, imagery, onomatopoeia, and personification to persuasively illustrate one or more central ideas they wish the reader to adopt. In order to be successful in a standardized reading comprehension test situation, a reader needs to be well acquainted in recognizing rhetoric and rhetorical devices.

Identifying Elements of Style

A writer's style is unique. The combinations of elements are carefully designed to create an effect on the reader. For example, the novels of J.K. Rowling are very different in style than the novels of Stephen King, yet both are designed to tell a compelling tale and to entertain readers. Furthermore, the articles found in *National Geographic* are vastly different from those a reader may encounter in *People* magazine, yet both have the same objective: to inform the reader. The difference is in the elements of style.

While there are many elements of style an author can employ, it's important to look at three things: the words they choose to use, the voice an author selects, and the fluency of sentence structure. Word choice is critical in persuasive or pictorial writing. While effective authors will choose words that are succinct, different authors will choose various words based on what they are trying to accomplish. For example, a reader would not expect to encounter the same words in a gothic novel that they would read in a scholastic article on gene therapy. An author whose intent is to paint a picture of a foreboding scene, will choose different words than an author who wants to persuade the reader that a particular political party has the most sound, ideological platform. A romance novelist will sound very different than a true crime writer.

The voice an author selects is also important to note. An author's voice is that element of style that indicates their personality. It's important that authors move us as readers; therefore, they will choose a voice that helps them do that. An author's voice may be satirical or authoritative. It may be light-hearted or serious in tone. It may be silly or humorous as well. Voice, as an element of style, can be vague in nature and difficult to identify, since it's also referred to as an author's tone, but it is that element unique to the author. It is the author's "self." A reader can expect an author's voice to vary across literary genres. A non-fiction author will generally employ a more neutral voice than an author of fiction, but use caution when trying to identify voice. Do not confuse an author's voice with a particular character's voice.

Another critical element of style involves how an author structures their sentences. An effective writer—one who wants to paint a vivid picture or strongly illustrate a central idea—will use a variety of sentence structures and sentence lengths. A reader is more likely to be confused if an author uses

choppy, unrelated sentences. Similarly, a reader will become bored and lose interest if an author repeatedly uses the same sentence structure. Good writing is fluent. It flows. Varying sentence structure keeps a reader engaged and helps reading comprehension. Consider the following example:

> The morning started off early. It was bright out. It was just daylight. The moon was still in the sky. He was tired from his sleepless night.

Then consider this text:

> Morning hit hard. He didn't remember the last time light hurt this bad. Sleep had been absent, and the very thought of moving towards the new day seemed like a hurdle he couldn't overcome.

Note the variety in sentence structure. The second passage is more interesting to read because the sentence fluency is more effective. Both passages paint the picture of a central character's reaction to dawn, but the second passage is more effective because it uses a variety of sentences and is more fluent than the first.

Elements of style can also include more recognizable components such as a story's setting, the type of narrative an author chooses, the mood they set, and the character conflicts employed. The ability to effectively understand the use of rhetoric demands the reader take note of an author's word choices, writing voice, and the ease of fluency employed to persuade, entertain, illustrate, or otherwise captivate a reader.

Identifying Rhetorical Devices

If one feels strongly about a subject, or has a passion for it, they choose strong words and phrases. Think of the types of rhetoric (or language) our politicians use. Each word, phrase, and idea is carefully crafted to elicit a response. Hopefully, that response is one of agreement to a certain point of view, especially among voters. Authors use the same types of language to achieve the same results. For example, the word "bad" has a certain connotation, but the words "horrid," "repugnant," and "abhorrent" paint a far better picture for the reader. They're more precise. They're interesting to read and they should all illicit stronger feelings in the reader than the word "bad." An author generally uses other devices beyond mere word choice to persuade, convince, entertain, or otherwise engage a reader.

Rhetorical devices are those elements an author utilizes in painting sensory, and hopefully persuasive ideas to which a reader can relate. They are numerable. Test takers will likely encounter one or more standardized test questions addressing varying rhetorical devices. This study guide will address the more common types: alliteration, irony, metaphor, simile, hyperbole, allegory, imagery, onomatopoeia, and personification, providing examples of each.

Alliteration is a device that uses repetitive beginning sounds in words to appeal to the reader. Classic tongue twisters are a great example of alliteration. *She sells sea shells down by the sea shore* is an extreme example of alliteration. Authors will use alliterative devices to capture a reader's attention. It's interesting to note that marketing also utilizes alliteration in the same way. A reader will likely remember products that have the brand name and item starting with the same letter. Similarly, many songs, poems, and catchy phrases use this device. It's memorable. Use of alliteration draws a reader's attention to ideas that an author wants to highlight.

Irony is a device that authors use when pitting two contrasting items or ideas against each other in order to create an effect. It's frequently used when an author wants to employ humor or convey a sarcastic tone. Additionally, it's often used in fictional works to build tension between characters, or between a particular character and the reader. An author may use *verbal irony* (sarcasm), *situational irony* (where actions or events have the opposite effect than what's expected), and *dramatic irony* (where the reader knows something a character does not). Examples of irony include:

- Dramatic Irony: An author describing the presence of a hidden killer in a murder mystery, unbeknownst to the characters but known to the reader.

- Situational Irony: An author relating the tale of a fire captain who loses her home in a five-alarm conflagration.

- Verbal Irony: This is where an author or character says one thing but means another. For example, telling a police officer "Thanks a lot" after receiving a ticket.

Metaphor is a device that uses a figure of speech to paint a visual picture of something that is not literally applicable. Authors relate strong images to readers, and evoke similar strong feelings using metaphors. Most often, authors will mention one thing in comparison to another more familiar to the reader. It's important to note that metaphors do not use the comparative words "like" or "as." At times, metaphors encompass common phrases such as clichés. At other times, authors may use mixed metaphors in making identification between two dissimilar things. Examples of metaphors include:

- An author describing a character's anger as *a flaming sheet of fire.*
- An author relating a politician as having been a folding chair under close questioning.
- A novel's character telling another character to *take a flying hike.*
- Shakespeare's assertion that *all the world's a stage.*

Simile is a device that compares two dissimilar things using the words "like" and "as." When using similes, an author tries to catch a reader's attention and use comparison of unlike items to make a point. Similes are commonly used and often develop into figures of speech and catch phrases.

Examples of similes include:

- An author describing a character as having a complexion like a faded lily.

- An investigative journalist describing his interview subject as being like cold steel and with a demeanor hard as ice.

- An author asserting the current political arena is just like a three-ring circus and as dry as day old bread.

Similes and metaphors can be confusing. When utilizing simile, an author will state one thing is like another. A metaphor states one thing is another. An example of the difference would be if an author states a character is *just like a fierce tiger and twice as angry,* as opposed to stating the character *is a fierce tiger and twice as angry.*

Hyperbole is simply an exaggeration that is not taken literally. A potential test taker will have heard or employed hyperbole in daily speech, as it is a common device we all use. Authors will use hyperbole to

draw a reader's eye toward important points and to illicit strong emotional and relatable responses. Examples of hyperbole include:

- An author describing a character as being as big as a house and twice the circumference of a city block.

- An author stating the city's water problem as being old as the hills and more expensive than a king's ransom in spent tax dollars.

- A journalist stating the mayoral candidate died of embarrassment when her tax records were made public.

Allegories are stories or poems with hidden meanings, usually a political or moral one. Authors will frequently use allegory when leading the reader to a conclusion. Allegories are similar to parables, symbols, and analogies. Often, an author will employ the use of allegory to make political, historical, moral, or social observations. As an example, Jonathan Swift's work *Gulliver's Travels into Several Remote Nations of the World* is an allegory in and of itself. The work is a political allegory of England during Jonathan Swift's lifetime. Set in the travel journal style plot of a giant amongst smaller people, and a smaller Gulliver amongst the larger, it is a commentary on Swift's political stance of existing issues of his age. Many fictional works are entire allegories in and of themselves. George Orwell's *Animal Farm* is a story of animals that conquer man and form their own farm society with swine at the top; however, it is not a literal story in any sense. It's Orwell's political allegory of Russian society during and after the Communist revolution of 1917. Other examples of allegory in popular culture include:

- Aesop's fable "The Tortoise and the Hare," which teaches readers that being steady is more important than being fast and impulsive.

- The popular *Hunger Games* by Suzanne Collins that teaches readers that media can numb society to what is truly real and important.

- Dr. Seuss's *Yertle the Turtle* which is a warning against totalitarianism and, at the time it was written, against the despotic rule of Adolf Hitler.

Imagery is a rhetorical device that an author employs when they use visual, or descriptive, language to evoke a reader's emotion. Use of imagery as a rhetorical device is broader in scope than this study guide addresses, but in general, the function of imagery is to create a vibrant scene in the reader's imagination and, in turn, tease the reader's ability to identify through strong emotion and sensory experience. In the simplest of terms, imagery, as a rhetoric device, beautifies literature.

An example of poetic imagery is below:

Pain has an element of blank

It cannot recollect

When it began, or if there were

A day when it was not.

It has no future but itself,

Its infinite realms contain

Its past, enlightened to perceive

New periods of pain.

In the above poem, Emily Dickenson uses strong imagery. Pain is equivalent to an "element of blank" or of nothingness. Pain cannot recollect a beginning or end, as if it was a person (see *personification* below). Dickenson appeals to the reader's sense of a painful experience by discussing the unlikelihood that discomfort sees a future, but does visualize a past and present. She simply indicates that pain, through the use of imagery, is cyclical and never ending. Dickenson's theme is one of painful depression and it is through the use of imagery that she conveys this to her readers.

Onomatopoeia is the author's use of words that create sound. Words like *pop* and *sizzle* are examples of onomatopoeia. When an author wants to draw a reader's attention in an auditory sense, they will use onomatopoeia. An author may also use onomatopoeia to create sounds as interjection or commentary. Examples include:

- An author describing a cat's vocalization as the kitten's chirrup echoed throughout the empty cabin.
- A description of a campfire as crackling and whining against its burning green wood.
- An author relating the sound of a car accident as *metallic screeching against crunching asphalt*.
- A description of an animal roadblock as being *a symphonic melody of groans, baas, and moans*.

Personification is a rhetorical device that an author uses to attribute human qualities to inanimate objects or animals. Once again, this device is useful when an author wants the reader to strongly relate to an idea. As in the example of George Orwell's *Animal Farm*, many of the animals are given the human abilities to speak, reason, apply logic, and otherwise interact as humans do. This helps the reader see how easily it is for any society to segregate into the haves and the have-nots through the manipulation of power. Personification is a device that enables the reader to empathize through human experience.

Examples of personification include:

- An author describing the wind as *whispering through the trees*.

- A description of a stone wall as being a hardened, unmovable creature made of cement and brick.

- An author attributing a city building as having slit eyes and an unapproachable, foreboding façade.

- An author describing spring as a beautiful bride, blooming in white, ready for summer's matrimony.

When identifying rhetorical devices, look for words and phrases that capture one's attention. Make note of the author's use of comparison between the inanimate and the animate. Consider words that make the reader feel sounds and envision imagery. Pay attention to the rhythm of fluid sentences and to the use of words that evoke emotion. The ability to identify rhetorical devices is another step in achieving successful reading comprehension and in being able to correctly answer standardized questions related to those devices.

Synthesis

Synthesis in reading involves the ability to fully comprehend text passages, and then going further by making new connections to see things in a new or different way. It involves a full thought process and requires readers to change the way they think about what they read. The PSAT/NMSQT will require a test taker to integrate new information that he or she already knows, and demonstrate an ability to express new thoughts.

Synthesis goes further than summary. When summarizing, a reader collects all of the information an author presents in a text passage, and restates it in an effective manner. Synthesis requires that the test taker not only summarize reading material, but be able to express new ideas based on the author's message. It is a full culmination of all reading comprehension strategies. It will require the test taker to order, recount, summarize, and recreate information into a whole new idea.

In utilizing synthesis, a reader must be able to form mental images about what they read, recall any background information they have about the topic, ask critical questions about the material, determine the importance of points an author makes, make inferences based on the reading, and finally be able to form new ideas based on all of the above skills. Synthesis requires the reader to make connections, visualize concepts, determine their importance, ask questions, make inferences, then fully synthesize all of this information into new thought.

Making Connections in Reading

There are three helpful thinking strategies to keep in mind when attempting to synthesize text passages:

- Think about how the content of a passage relates to life experience.
- Think about how the content of a passage relates to other text.
- Think about how the content of a passage relates to the world in general.

When reading a given passage, the test taker should actively think about how the content relates to their life experience. While the author's message may express an opinion different from what the reader believes, or express ideas with which the reader is unfamiliar, a good reader will try to relate any of the author's details to their own familiar ground. A reader should use context clues to understand unfamiliar terminology, and recognize familiar information they have encountered in prior experience. Bringing prior life experience and knowledge to the test-taking situation is helpful in making connections. The ability to relate an unfamiliar idea to something the reader already knows is critical in understanding unique and new ideas.

When trying to make connections while reading, keep the following questions in mind:

- How does this feel familiar in personal experience?
- How is this similar to or different from other reading?
- How is this familiar in the real world?
- How does this relate to the world in general?

A reader should ask themselves these questions during the act of reading in order to actively make connections to past and present experiences. Utilizing the ability to make connections is an important step in achieving synthesis.

Determining Importance in Reading

Being able to determine what is most important while reading is critical to synthesis. It is the difference between being able to tell what is necessary to full comprehension and that which is interesting but not necessary.

When determining the importance of an author's ideas, consider the following:

- Ask how critical an author's particular idea, assertion, or concept is to the overall message.

- Ask "is this an interesting fact or is this information essential to understanding the author's main idea?"

- Make a simple chart. On one side, list all of the important, essential points an author makes and on the other, list all of the interesting yet non-critical ideas.

- Highlight, circle, or underline any dates or data in non-fiction passages. Pay attention to headings, captions, and any graphs or diagrams.

- When reading a fictional passage, delineate important information such as theme, character, setting, conflict (what the problem is), and resolution (how the problem is fixed). Most often, these are the most important aspects contained in fictional text.

- If a non-fiction passage is instructional in nature, take physical note of any steps in the order of their importance as presented by the author. Look for words such as *first*, *next*, *then*, and *last*.

Determining the importance of an author's ideas is critical to synthesis in that it requires the test taker to parse out any unnecessary information and demonstrate they have the ability to make sound determination on what is important to the author, and what is merely a supporting or less critical detail.

Asking Questions While Reading

A reader must ask questions while reading. This demonstrates their ability to critically approach information and apply higher thinking skills to an author's content. Some of these questions have been addressed earlier in this section. A reader must ask what is or isn't important, what relates to their experience, and what relates to the world in general? However, it's important to ask other questions as well in order to make connections and synthesize reading material. Consider the following partial list of possibilities:

- What type of passage is this? Is it fiction? Non-fiction? Does it include data?

- Based on the type of passage, what information should be noted in order to make connections, visualize details, and determine importance?

- What is the author's message or theme? What is it they want the reader to understand?

- Is this passage trying to convince readers of something? What is it? If so, is the argument logical, convincing, and effective? How so? If not, how not?

- What do readers already know about this topic? Are there other viewpoints that support or contradict it?

- Is the information in this passage current and up to date?

- Is the author trying to teach readers a lesson? If so, what is it? Is there a moral to this story?

- How does this passage relate to experience?

- What is not as understandable in this passage? What context clues can help with understanding?

- What conclusions can be drawn? What predictions can be made?

Again, the above should be considered only a small example of the possibilities. Any question the reader asks while reading will help achieve synthesis and full reading comprehension.

Analysis of History/Social Studies Excerpts

The PSAT/NMSQT will test for the ability to read substantial, historically based excerpts, and then answer comprehension questions based on content. The test taker will encounter at least two U.S. history, or social science, passages within the test. One is likely to be from a U.S. founding document or work that has had great impact on history. The test may also include one or more passages from social sciences such as economics, psychology, or sociology.

For these types of questions, the test taker will need to utilize all the reading comprehension skills discussed above, but mastery of further skills will help. This section addresses those skills.

Comprehending Test Questions Prior to Reading

While preparing for a historical passage on a standardized test, first read the test questions, and then quickly scan the test answers prior to reading the passage itself. Notice there is a difference between the terms *read* and *scans*. Reading involves full concentration while addressing every word. Scanning involves quickly glancing at text in chunks, noting important dates, words, and ideas along the way. Reading test questions will help the test taker know what information to focus on in the historical passage. Scanning answers will help the test taker focus on possible answer options while reading the passage.

When reading standardized test questions that address historical passages, be sure to clearly understand what each question is asking. Is a question asking about vocabulary? Is another asking for the test taker to find a specific historical fact? Do any of the questions require the test taker to draw conclusions, identify an author's topic, tone, or position? Knowing what content to address will help the test taker focus on the information they will be asked about later. However, the test taker should approach this reading comprehension technique with some caution. It is tempting to only look for the right answers within any given passage. Do not put on "reading blinders" and ignore all other information presented in a passage. It is important to fully read every passage and not just scan it. Strictly looking for what may be the right answers to test questions can cause the test taker to ignore important contextual clues that actually require critical thinking in order to identify correct answers. Scanning a passage for what appears to be wrong answers can have a similar result.

When reading test questions prior to tackling a historical passage, be sure to understand what skills the test is assessing, and then fully read the related passage with those skills in mind. Focus on every word in both the test questions and the passage itself. Read with a critical eye and a logical mind.

Reading for Factual Information

Standardized test questions that ask for factual information are usually straightforward. These types of questions will either ask the test taker to confirm a fact by choosing a correct answer, or to select a correct answer based on a negative fact question.

For example, the test taker may encounter a passage from Lincoln's Gettysburg address. A corresponding test question may ask the following:

> Which war is Abraham Lincoln referring to in the following passage?: "Now we are engaged in a great civil war, testing whether that nation, or any nation so conceived and so dedicated, can long endure."

This type of question is asking the test taker to confirm a simple fact. Given options such as World War I, the War of Spanish Succession, World War II, and the American Civil War, the test taker should be able to correctly identify the American Civil War based on the words "civil war" within the passage itself, and, hopefully, through general knowledge. In this case, reading the test question and scanning answer options ahead of reading the Gettysburg address would help quickly identify the correct answer. Similarly, a test taker may be asked to confirm a historical fact based on a negative fact question. For example, a passage's corresponding test question may ask the following:

> Which option is incorrect based on the above passage?

Given a variety of choices speaking about which war Abraham Lincoln was addressing, the test taker would need to eliminate all correct answers pertaining to the American Civil War and choose the answer choice referencing a different war. In other words, the correct answer is the one that contradicts the information in the passage.

It is important to remember that reading for factual information is straightforward. The test taker must distinguish fact from bias. Factual statements can be proven or disproven independent of the author and from a variety of other sources. Remember, successfully answering questions regarding factual information may require the test taker to re-read the passage, as these types of questions test for attention to detail.

Reading for Tone, Message, and Effect

The PSAT/NMSQT does not just address a test taker's ability to find facts within a historical reading passage; it also determines a reader's ability to determine an author's viewpoint through the use of tone, message, and overall effect. This type of reading comprehension requires inference skills, deductive reasoning skills, the ability to draw logical conclusions, and overall critical thinking skills. Reading for factual information is straightforward. Reading for an author's tone, message, and overall effect is not. It's key to read carefully when asked test questions that address a test taker's ability to these writing devices. These are not questions that can be easily answered by quickly scanning for the right information.

Tone

An author's *tone* is the use of particular words, phrases, and writing style to convey an overall meaning. Tone expresses the author's attitude towards a particular topic. For example, a historical reading passage may begin like the following:

> The presidential election of 1960 ushered in a new era, a new Camelot, a new phase of forward thinking in U.S. politics that embraced brash action, unrest, and responded with admirable leadership.

From this opening statement, a reader can draw some conclusions about the author's attitude towards President John F. Kennedy. Furthermore, the reader can make additional, educated guesses about the state of the Union during the 1960 presidential election. By close reading, the test taker can determine that the repeated use of the word *new* and words such as *admirable leadership* indicate the author's tone of admiration regarding the President's boldness. In addition, the author assesses that the era during President Kennedy's administration was problematic through the use of the words *brash action* and *unrest.* Therefore, if a test taker encountered a test question asking about the author's use of tone and their assessment of the Kennedy administration, the test taker should be able to identify an answer indicating admiration. Similarly, if asked about the state of the Union during the 1960s, a test taker should be able to correctly identify an answer indicating political unrest.

When identifying an author's tone, the following list of words may be helpful. This is not an inclusive list. Generally, parts of speech that indicate attitude will also indicate tone:

- Comical
- Angry
- Ambivalent
- Scary
- Lyrical
- Matter-of-fact
- Judgmental
- Sarcastic
- Malicious
- Objective
- Pessimistic
- Patronizing
- Gloomy
- Instructional
- Satirical
- Formal
- Casual

Message

An author's *message* is the same as the overall meaning of a passage. It is the main idea, or the main concept the author wishes to convey. An author's message may be stated outright or it may be implied. Regardless, the test taker will need to use careful reading skills to identify an author's message or purpose.

Often, the message of a particular passage can be determined by thinking about why the author wrote the information. Many historical passages are written to inform and to teach readers established, factual information. However, many historical works are also written to convey biased ideas to readers. Gleaning bias from an author's message in a historical passage can be difficult, especially if the reader is presented with a variety of established facts as well. Readers tend to accept historical writing as factual. This is not always the case. Any discerning reader who has tackled historical information on topics such as United States political party agendas can attest that two or more works on the same topic may have completely different messages supporting or refuting the value of the identical policies. Therefore, it is important to critically assess an author's message separate from factual information. One author, for example, may point to the rise of unorthodox political candidates in an election year based on the failures of the political party in office while another may point to the rise of the same candidates in the same election year based on the current party's successes. The historical facts of what has occurred leading up to an election year are not in refute. Labeling those facts as a failure or a success is a bias within an author's overall *message*, as is excluding factual information in order to further a particular point. In a standardized testing situation, a reader must be able to critically assess what the author is trying to say separate from the historical facts that surround their message.

Using the example of Lincoln's Gettysburg Address, a test question may ask the following:

> What is the message the author is trying to convey through this address?

Then they will ask the test taker to select an answer that best expresses Lincoln's *message* to his audience. Based on the options given, a test taker should be able to select the answer expressing the idea that Lincoln's audience should recognize the efforts of those who died in the war as a sacrifice to preserving human equality and self-government.

Effect

The *effect* an author wants to convey is when an author wants to impart a particular mood in their message. An author may want to challenge a reader's intellect, inspire imagination, or spur emotion. An author may present information to appeal to a physical, aesthetic, or transformational sense. Take the following text as an example:

> In 1963, Martin Luther King stated "I have a dream." The gathering at the Lincoln Memorial was the beginning of the Civil Rights movement and, with its reference to the Emancipation Proclamation, Dr. King's words electrified those who wanted freedom and equality while rising from hatred and slavery. It was the beginning of radical change.

The test taker may be asked about the effect this statement might have on King's audience. Through careful reading of the passage, the test taker should be able to choose an answer that best identifies an effect of grabbing the audience's attention. The historical facts are in place: King made the speech in 1963 at the Lincoln Memorial, kicked off the civil rights movement, and referenced the Emancipation Proclamation. The words *electrified* and *radical change* indicate the effect the author wants the reader to understand as a result of King's speech. In this historical passage, facts are facts. However, the author's message goes above the facts to indicate the effect the message had on the audience and, in addition, the effect the event should have on the reader.

When reading historical passages, the test taker should perform due diligence in their awareness of the test questions and answers up front. From there, the test taker should carefully, and critically, read all

historical excerpts with an eye for detail, tone, message (biased or unbiased), and effect. Being able to synthesize these skills will result in success in a standardized testing situation.

Analysis of Science Excerpts

The PSAT/NMSQT includes at least two science passages that address the fundamental concepts of Earth science, biology, chemistry, and/or physics. While prior general knowledge of these subjects is helpful in determining correct test answers, the test taker's ability to comprehend the passages is key to success. When reading scientific excerpts, the test taker must be able to examine quantitative information, identify hypotheses, interpret data, and consider implications of the material they are presented with. It is helpful, at this point, to reference the above section on comprehending test questions prior to reading. The same rules apply: read questions and scan questions, along with their answers, prior to fully reading a passage. Be informed prior to approaching a scientific text. A test taker should know what they will be asked and how to apply their reading skills. In this section of the test, it is also likely that a test taker will encounter graphs and charts to assess their ability to interpret scientific data with an appropriate conclusion. This section will determine the skills necessary to address scientific data presented through identifying hypotheses, through reading and examining data, and through interpreting data representation passages.

Examine Hypotheses

When presented with fundamental, scientific concepts, it is important to read for understanding. The most basic skill in achieving this literacy is to understand the concept of hypothesis and moreover, to be able to identify it in a particular passage. A hypothesis is a proposed idea that needs further investigation in order to be proven true or false. While it can be considered an educated guess, a hypothesis goes more in depth in its attempt to explain something that is not currently accepted within scientific theory. It requires further experimentation and data gathering to test its validity and is subject to change, based on scientifically conducted test results. Being able to read a science passage and understand its main purpose, including any hypotheses, helps the test taker understand data-driven evidence. It helps the test taker to be able to correctly answer questions about the science excerpt they are asked to read.

When reading to identify a hypothesis, a test taker should ask, "What is the passage trying to establish? What is the passage's main idea? What evidence does the passage contain that either supports or refutes this idea?" Asking oneself these questions will help identify a hypothesis. Additionally, hypotheses are logical statements that are testable and use very precise language.

Review the following hypothesis example:

> Consuming excess sugar in the form of beverages has a greater impact on childhood obesity and subsequent weight gain than excessive sugar from food.

While this is likely a true statement, it is still only a conceptual idea in a text passage regarding sugar consumption in childhood obesity, unless the passage also contains tested data that either proves or disproves the statement. A test taker could expect the rest of the passage to cite data proving that children who drink empty calories and don't exercise will, in fact, be obese.

A hypothesis goes further in that, given its ability to be proven or disproven, it may result in further hypotheses that require extended research. For example, the hypothesis regarding sugar consumption in drinks, after undergoing rigorous testing, may lead scientists to state another hypothesis such as the following:

> Consuming excess sugar in the form of beverages as opposed to food items is a habit found in mostly sedentary children.

This new, working hypothesis further focuses not just on the source of an excess of calories, but tries an "educated guess" that empty caloric intake has a direct, subsequent impact on physical behavior.

The data-driven chart below is similar to an illustration a test taker might see in relation to the hypothesis on sugar consumption in children:

Behaviors of Healthy and Unhealthy Kids

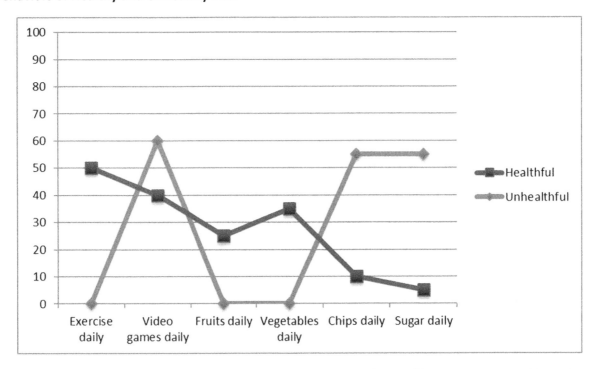

While this guide will address other data-driven passages a test taker could expect to see within a given science excerpt, note that the hypothesis regarding childhood sugar intake and rate of exercise has undergone scientific examination and yielded results that support its truth.

When reading a science passage to determine its hypothesis, a test taker should look for a concept that attempts to explain a phenomenon, is testable, logical, precisely worded, and yields data-driven results. The test taker should scan the presented passage for any word or data-driven clues that will help identify the hypothesis, and then be able to correctly answer test questions regarding the hypothesis based on their critical thinking skills.

Interpreting Data and Considering Implications

The PSAT/NMSQT is likely to contain one or more data-driven science passages that require the test taker to examine evidence within a particular type of graphic. The test taker will then be required to interpret the data and answer questions demonstrating their ability to draw logical conclusions.

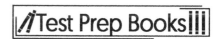

In general, there are two types of data: qualitative and quantitative. Science passages may contain both, but simply put, quantitative data is reflected numerically and qualitative is not. Qualitative data is based on its qualities. In other words, qualitative data tends to present information more in subjective generalities (for example, relating to size or appearance). Quantitative data is based on numerical findings such as percentages. Quantitative data will be described in numerical terms. While both types of data are valid, the test taker will more likely be faced with having to interpret quantitative data through one or more graphic(s), and then be required to answer questions regarding the numerical data. The section of this study guide briefly addresses how data may be displayed in line graphs, bar charts, circle graphs, and scatter plots. A test taker should take the time to learn the skills it takes to interpret quantitative data. An example of a line graph is as follows:

Cell Phone Use in Kiteville, 2000-2006

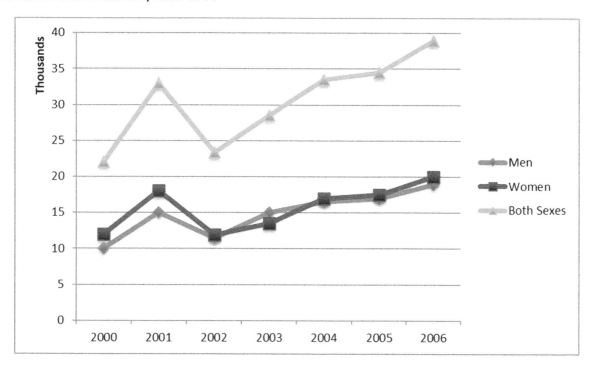

A line graph presents quantitative data on both horizontal (side to side) and vertical (up and down) axes. It requires the test taker to examine information across varying data points. When reading a line graph, a test taker should pay attention to any headings, as these indicate a title for the data it contains. In the above example, the test taker can anticipate the line graph contains numerical data regarding the use of cellphones during a certain time period. From there, a test taker should carefully read any outlying words or phrases that will help determine the meaning of data within the horizontal and vertical axes. In this example, the vertical axis displays the total number of people in increments of 5,000. Horizontally, the graph displays yearly markers, and the reader can assume the data presented accounts for a full calendar year. In addition, the line graph also defines its data points by shapes. Some data points represent the number of men. Some data points represent the number of women, and a third type of data point represents the number of both sexes combined.

A test taker may be asked to read and interpret the graph's data, then answer questions about it. For example, the test may ask, *In which year did men seem to decrease cellphone use?* then require the test taker to select the correct answer. Similarly, the test taker may encounter a question such as *Which year yielded the highest number of cellphone users overall?* The test taker should be able to identify the correct answer as 2006.

A **bar graph** presents quantitative data through the use of lines or rectangles. The height and length of these lines or rectangles corresponds to the magnitude of the numerical data for that particular category or attribute. The data presented may represent information over time, showing shaded data over time or over other defined parameters. A bar graph will also utilize horizontal and vertical axes. An example of a bar graph is as follows:

Population Growth in Major U.S. Cities

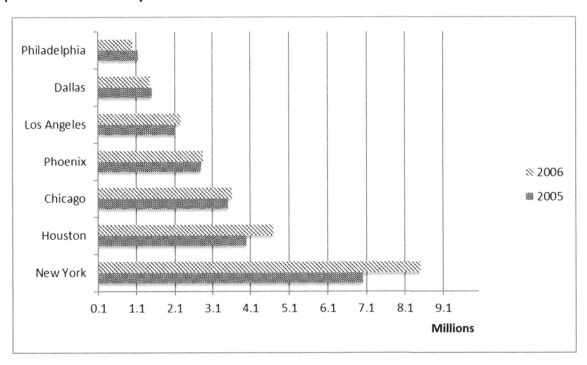

Reading the data in a bar graph is similar to the skills needed to read a line graph. The test taker should read and comprehend all heading information, as well as information provided along the horizontal and vertical axes. Note that the graph pertains to the population of some major U.S. cities. The "values" of these cities can be found along the left side of the graph, along the vertical axis. The population values can be found along the horizontal axes. Notice how the graph uses shaded bars to depict the change in population over time, as the heading indicates. Therefore, when the test taker is asked a question such as, *Which major U.S. city experienced the greatest amount of population growth during the depicted two year cycle,* the reader should be able to determine a correct answer of New York. It is important to pay particular attention to color, length, data points, and both axes, as well as any outlying header information in order to be able to answer graph-like test questions.

A circle graph presents quantitative data in the form of a circle (also sometimes referred to as a pie chart). The same principles apply: the test taker should look for numerical data within the confines of the circle itself but also note any outlying information that may be included in a header, footer, or to the side of the circle. A circle graph will not depict horizontal or vertical axis information, but will instead

rely on the reader's ability to visually take note of segmented circle pieces and apply information accordingly. An example of a circle graph is as follows:

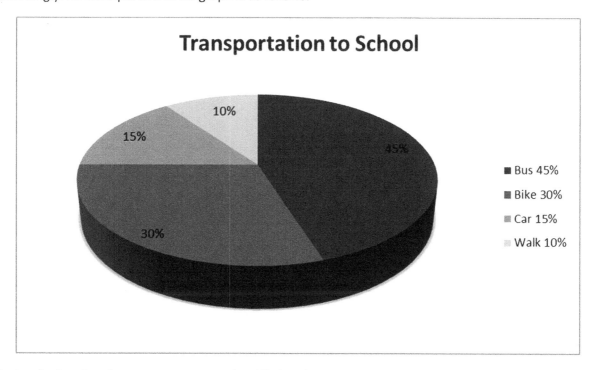

Notice the heading "Transportation to School." This should indicate to the test taker that the topic of the circle graph is how people traditionally get to school. To the right of the graph, the reader should comprehend that the data percentages contained within it directly correspond to the method of transportation. In this graph, the data is represented through the use shades and pattern. Each transportation method has its own shade. For example, if the test taker was then asked, *Which method of school transportation is most widely utilized,* the reader should be able to identify school bus as the correct answer.

Be wary of test questions that ask test takers to draw conclusions based on information that is not present. For example, it is not possible to determine, given the parameters of this circle graph, whether the population presented is of a particular gender or ethnic group. This graph does not represent data from a particular city or school district. It does not distinguish between student grade levels and, although the reader could infer that the typical student must be of driving age if cars are included, this is not necessarily the case. Elementary school students may rely on parents or others to drive them by personal methods. Therefore, do not read too much into data that is not presented. Only rely on the quantitative data that is presented in order to answer questions.

A scatter plot or scatter diagram is a graph that depicts quantitative data across plotted points. It will involve at least two sets of data. It will also involve horizontal and vertical axes.

An example of a scatter plot is as follows:

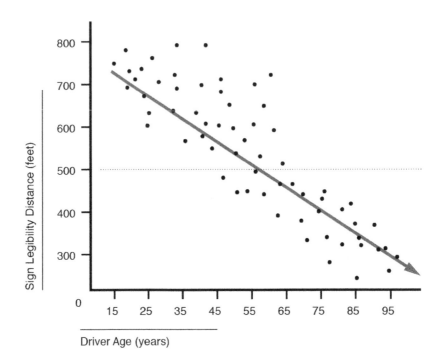

The skills needed to address a scatter plot are essentially the same as in other graph examples. Note any topic headings, as well as horizontal or vertical axis information. In the sample above, the reader can determine the data addresses a driver's ability to correctly and legibly read road signs as related to their age. Again, note the information that is absent. The test taker is not given the data to assess a time period, location, or driver gender. It simply requires the reader to note an approximate age to the ability to correctly identify road signs from a distance measured in feet. Notice that the overall graph also displays a trend. In this case, the data indicates a negative one and possibly supports the hypothesis that as a driver ages, their ability to correctly read a road sign at over 500 feet tends to decline over time. If the test taker were to be asked, *At what approximation in feet does a sixteen-year-old driver correctly see and read a street sign,* the answer would be the option closest to 700 feet.

Reading and examining scientific data in excerpts involves all of a reader's contextual reading, data interpretation, drawing logical conclusions based only on the information presented, and their application of critical thinking skills across a set of interpretive questions. Thorough comprehension and attention to detail is necessary to achieve test success.

Practice Questions

Questions 1–5 are based on the following passage:

"Mademoiselle Eugénie is pretty—I think I remember that to be her name."

"Very pretty, or rather, very beautiful," replied Albert, "but of that style of beauty which I don't appreciate; I am an ungrateful fellow."

"Really," said Monte Cristo, lowering his voice, "you don't appear to me to be very enthusiastic on the subject of this marriage."

"Mademoiselle Danglars is too rich for me," replied Morcerf, "and that frightens me."

"Bah," exclaimed Monte Cristo, "that's a fine reason to give. Are you not rich yourself?"

"My father's income is about 50,000 francs per annum; and he will give me, perhaps, ten or twelve thousand when I marry."

"That, perhaps, might not be considered a large sum, in Paris especially," said the count; "but everything doesn't depend on wealth, and it's a fine thing to have a good name, and to occupy a high station in society. Your name is celebrated, your position magnificent; and then the Comte de Morcerf is a soldier, and it's pleasing to see the integrity of a Bayard united to the poverty of a Duguesclin; disinterestedness is the brightest ray in which a noble sword can shine. As for me, I consider the union with Mademoiselle Danglars a most suitable one; she will enrich you, and you will ennoble her."

Albert shook his head, and looked thoughtful. "There is still something else," said he.

"I confess," observed Monte Cristo, "that I have some difficulty in comprehending your objection to a young lady who is both rich and beautiful."

"Oh," said Morcerf, "this repugnance, if repugnance it may be called, isn't all on my side."

"Whence can it arise, then? for you told me your father desired the marriage."

"It's my mother who dissents; she has a clear and penetrating judgment, and doesn't smile on the proposed union. I cannot account for it, but she seems to entertain some prejudice against the Danglars."

"Ah," said the count, in a somewhat forced tone, "that may be easily explained; the Comtesse de Morcerf, who is aristocracy and refinement itself, doesn't relish the idea of being allied by your marriage with one of ignoble birth; that is natural enough."

Excerpt from the Count of Monte Cristo by Alexandre Dumas

1. The meaning of the word *repugnance* is closest to:
   a. Strong resemblance
   b. Strong dislike
   c. Extreme shyness
   d. Extreme dissimilarity

2. What is Albert's attitude towards his impending marriage?
   a. Pragmatic
   b. Romantic
   c. Indifferent
   d. Apprehensive

3. Which sentence is true of Albert's mother?
   a. She belongs to a noble family.
   b. She often makes poor choices.
   c. She is primarily occupied with money.
   d. She is unconcerned about her son's future.

4. Why is the Count puzzled by Albert's attitude toward his marriage?
   a. He seems reluctant to marry Eugénie, despite her wealth and beauty.
   b. He is marrying against his father's wishes, despite usually following his advice.
   c. He appears excited to marry someone he doesn't love, despite being a hopeless romantic.
   d. He expresses reverence towards Eugénie, despite being from a higher social class than her.

5. What does the word *ennoble* mean in the middle of the passage?
   a. To create beauty in another person
   b. To endow someone with wealth
   c. To make someone chaste again
   d. To give someone a noble rank or title

Answer Explanations

1. B: Strong dislike. This vocabulary question can be answered using context clues. Based on the rest of the conversation, the reader can gather that Albert isn't looking forward to his marriage. As the Count notes that "you don't appear to me to be very enthusiastic on the subject of this marriage," and also remarks on Albert's "objection to a young lady who is both rich and beautiful," readers can guess Albert's feelings. The answer choice that most closely matches "objection" and "not . . . very enthusiastic" is *B, strong dislike*.

2. D: Apprehensive. As in question 7, there are many clues in the passage that indicate Albert's attitude towards his marriage—far from enthusiastic, he has many reservations. This question requires test takers to understand the vocabulary in the answer choices. *Pragmatic* is closest in meaning to *realistic*, and *indifferent* means *uninterested*. The only word related to feeling worried, uncertain, or unfavorable about the future is *apprehensive*.

3. A: She belongs to a noble family. Though Albert's mother doesn't appear in the scene, there's more than enough information to answer this question. More than once is his family's noble background mentioned (not to mention that Albert's mother is the Comtesse de Morcerf, a noble title). The other answer choices can be eliminated—she is obviously deeply concerned about her son's future; money isn't her highest priority because otherwise she would favor a marriage with the wealthy Danglars; and Albert describes her "clear and penetrating judgment," meaning she makes good decisions.

4. A: He seems reluctant to marry Eugénie, despite her wealth and beauty. This is a reading comprehension question, and the answer can be found in the following lines: "'I confess,' observed Monte Cristo, "that I have some difficulty in comprehending your objection to a young lady who is both rich and beautiful.'" Choice *B* is the opposite (Albert's father is the one who insists on the marriage), Choice *C* incorrectly represents Albert's eagerness to marry, and Choice *D* describes a more positive attitude than Albert actually feels (*repugnance*).

5. D: The meaning of the word *ennoble* in the middle of the paragraph means to give someone a noble rank or title. In the passage, we can infer that Albert is noble but not rich, and Mademoiselle Eugénie is rich but not noble.

Writing and Language Test

The PSAT Writing and Language Test contains a series of passages that must be read along with questions pertaining to each passage. The task is not to recall or restate a passage's content, but to analyze *how* the content is presented and answer questions about how to improve it.

Expression of Ideas

This test is about *how* the information is communicated rather than the subject matter itself. The good news is there isn't any writing! Instead, it's like being an editor helping the writer find the best ways to express their ideas. Things to consider include: how well a topic is developed, how accurately facts are presented, whether the writing flows logically and cohesively, and how effectively the writer uses language. This can seem like a lot to remember, but these concepts are the same ones taught way back in elementary school.

One last thing to remember while going through this guide is not to be intimidated by the terminology. Phrases like "pronoun-antecedent agreement" and "possessive determiners" can sound confusing and complicated, but the ideas are often quite simple and easy to understand. Though proper terminology is used to explain the rules and guidelines, the PSAT Writing and Language Test is not a technical grammar test.

Organization

Good writing is not merely a random collection of sentences. No matter how well written, sentences must relate and coordinate appropriately to one another. If not, the writing seems random, haphazard, and disorganized. Therefore, good writing must be *organized* (where each sentence fits a larger context and relates to the sentences around it).

Transition Words

The writer should act as a guide, showing the reader how all the sentences fit together. Consider this example:

> Seat belts save more lives than any other automobile safety feature. Many studies show that airbags save lives as well. Not all cars have airbags. Many older cars don't. Air bags aren't entirely reliable. Studies show that in 15% of accidents, airbags don't deploy as designed. Seat belt malfunctions are extremely rare.

There's nothing wrong with any of these sentences individually, but together they're disjointed and difficult to follow. The best way for the writer to communicate information is through the use of *transition words*. Here are examples of transition words and phrases that tie sentences together, enabling a more natural flow:

- To show causality: as a result, therefore, and consequently
- To compare and contrast: *however, but*, and *on the other hand*
- To introduce examples: *for instance, namely*, and *including*
- To show order of importance: *foremost, primarily, secondly*, and *lastly*

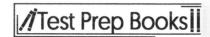

The above is not a complete list of transitions. There are many more that can be used; however, most f
into these or similar categories. The important point is that the words should clearly show the
relationship between sentences, supporting information, and the main idea.

Here is an update to the previous example using transition words. These changes make it easier to read
and bring clarity to the writer's points:

> Seat belts save more lives than any other automobile safety feature. Many studies show that
> airbags save lives as well. However, not all cars have airbags. For instance, some older cars
> don't. Furthermore, air bags aren't entirely reliable. For example, studies show that in 15% of
> accidents, airbags don't deploy as designed. But, on the other hand, seat belt malfunctions are
> extremely rare.

Also be prepared to analyze whether the writer is using the best transition word or phrase for the
situation. Take this sentence for example: "As a result, seat belt malfunctions are extremely rare." This
sentence doesn't make sense in the context above because the writer is trying to show the *contrast*
between seat belts and airbags, not the causality.

Logical Sequence

Even if the writer includes plenty of information to support their point, the writing is only effective when
the information is in a logical order. *Logical sequencing* is really just common sense, but it's an
important writing technique. First, the writer should introduce the main idea, whether for a paragraph,
a section, or the entire piece. Second, they should present evidence to support the main idea by using
transitional language. This shows the reader how the information relates to the main idea and to the
sentences around it. The writer should then take time to interpret the information, making sure
necessary connections are obvious to the reader. Finally, the writer can summarize the information in a
closing section.

Although most writing follows this pattern, it isn't a set rule. Sometimes writers change the order for
effect. For example, the writer can begin with a surprising piece of supporting information to grab the
reader's attention, and then transition to the main idea. Thus, if a passage doesn't follow the logical
order, don't immediately assume it's wrong. However, most writing usually settles into a logical
sequence after a nontraditional beginning.

Focus

Good writing stays *focused* and on topic. During the test, determine the main idea for each passage and
then look for times when the writer strays from the point they're trying to make. Let's go back to the
seat belt example. If the writer suddenly begins talking about how well airbags, crumple zones, or other
safety features work to save lives, they might be losing focus from the topic of "safety belts."

Focus can also refer to individual sentences. Sometimes the writer does address the main topic, but in a
confusing way. For example:

> Thanks to seat belt usage, survival in serious car accidents has shown a consistently steady
> increase since the development of the retractable seat belt in the 1950s.

This statement is definitely on topic, but it's not easy to follow. A simpler, more focused version of this sentence might look like this:

Seat belts have consistently prevented car fatalities since the 1950s.

Providing *adequate information* is another aspect of focused writing. Statements like "seat belts are important" and "many people drive cars" are true, but they're so general that they don't contribute much to the writer's case. When reading a passage, watch for these kinds of unfocused statements.

Introductions and Conclusions

Examining the writer's strategies for introductions and conclusions puts the reader in the right mindset to interpret the rest of the passage. Look for methods the writer might use for introductions such as:

- Stating the main point immediately, followed by outlining how the rest of the piece supports this claim.

- Establishing important, smaller pieces of the main idea first, and then grouping these points into a case for the main idea.

- Opening with a quotation, anecdote, question, seeming paradox, or other piece of interesting information, and then using it to lead to the main point.

Whatever method the writer chooses, the introduction should make their intention clear, establish their voice as a credible one, and encourage a person to continue reading.

Conclusions tend to follow a similar pattern. In them, the writer restates their main idea a final time, often after summarizing the smaller pieces of that idea. If the introduction uses a quote or anecdote to grab the reader's attention, the conclusion often makes reference to it again. Whatever way the writer chooses to arrange the conclusion, the final restatement of the main idea should be clear and simple for the reader to interpret.

Finally, conclusions shouldn't introduce any new information.

Precision

People often think of *precision* in terms of math, but precise word choice is another key to successful writing. Since language itself is imprecise, it's important for the writer to find the exact word or words to convey the full, intended meaning of a given situation. For example:

The number of deaths has gone down since seat belt laws started.

There are several problems with this sentence. First, the word *deaths* is too general. From the context, it's assumed that the writer is referring only to *deaths* caused by car accidents. However, without clarification, the sentence lacks impact and is probably untrue. The phrase "gone down" might be accurate, but a more precise word could provide more information and greater accuracy. Did the numbers show a slow and steady decrease of highway fatalities or a sudden drop? If the latter is true, the writer is missing a chance to make their point more dramatically. Instead of "gone down" they could substitute *plummeted*, *fallen drastically*, or *rapidly diminished* to bring the information to life. Also, the phrase "seat belt laws" is unclear. Does it refer to laws requiring cars to include seat belts or to laws requiring drivers and passengers to use them? Finally, *started* is not a strong verb. Words like *enacted* or

adopted are more direct and make the content more real. When put together, these changes create a far more powerful sentence:

> The number of highway fatalities has plummeted since laws requiring seat belt usage were enacted.

However, it's important to note that precise word choice can sometimes be taken too far. If the writer of the sentence above takes precision to an extreme, it might result in the following:

> The incidence of high-speed, automobile accident related fatalities has decreased 75% and continued to remain at historical lows since the initial set of federal legislations requiring seat belt use were enacted in 1992.

This sentence is extremely precise, but it takes so long to achieve that precision that it suffers from a lack of clarity. Precise writing is about finding the right balance between information and flow. This is also an issue of *conciseness* (discussed in the next section).

The last thing to consider with precision is a word choice that's not only unclear or uninteresting, but also confusing or misleading. For example:

> The number of highway fatalities has become hugely lower since laws requiring seat belt use were enacted.

In this case, the reader might be confused by the word *hugely*. Huge means large, but here the writer uses *hugely* to describe something small. Though most readers can decipher this, doing so disconnects them from the flow of the writing and makes the writer's point less effective.

On the test, there can be questions asking for alternatives to the writer's word choice. In answering these questions, always consider the context and look for a balance between precision and flow.

Conciseness

"Less is more" is a good rule to follow when writing a sentence. Unfortunately, writers often include extra words and phrases that seem necessary at the time, but add nothing to the main idea. This confuses the reader and creates unnecessary repetition. Writing that lacks *conciseness* is usually guilty of excessive wordiness and redundant phrases. Here's an example containing both of these issues:

> When legislators decided to begin creating legislation making it mandatory for automobile drivers and passengers to make use of seat belts while in cars, a large number of them made those laws for reasons that were political reasons.

There are several empty or "fluff" words here that take up too much space. These can be eliminated while still maintaining the writer's meaning. For example:

- "decided to begin" could be shortened to "began"
- "making it mandatory for" could be shortened to "requiring"
- "make use of" could be shortened to "use"
- "a large number" could be shortened to "many"

In addition, there are several examples of redundancy that can be eliminated:

- "legislators decided to begin creating legislation" and "made those laws"
- "automobile drivers and passengers" and "while in cars"
- "reasons that were political reasons"

These changes are incorporated as follows:

> When legislators began requiring drivers and passengers to use seat belts, many of them did so for political reasons.

There are many examples of redundant phrases, such as "add an additional," "complete and total," "time schedule," and "transportation vehicle." If asked to identify a redundant phrase on the test, look for words that are close together with the same (or similar) meanings.

Proposition

The *proposition* (also called the *claim* since it can be true or false) is a clear statement of the point or idea the writer is trying to make. The length or format of a proposition can vary, but it often takes the form of a *topic sentence*. A good topic sentence is:

- Clear: does not weave a complicated web of words for the reader to decode or unwrap

- Concise: presents only the information needed to make the claim and doesn't clutter up the statement with unnecessary details

- Precise: clarifies the exact point the writer wants to make and doesn't use broad, overreaching statements

Look at the following example:

> The civil rights movement, from its genesis in the Emancipation Proclamation to its current struggles with de facto discrimination, has changed the face of the United States more than any other factor in its history.

Is the statement clear? Yes, the statement is fairly clear, although other words can be substituted for "genesis" and "de facto" to make it easier to understand.

Is the statement concise? No, the statement is not concise. Details about the Emancipation Proclamation and the current state of the movement are unnecessary for a topic sentence. Those details should be saved for the body of the text.

Is the statement precise? No, the statement is not precise. What exactly does the writer mean by "changed the face of the United States"? The writer should be more specific about the effects of the movement. Also, suggesting that something has a greater impact than anything else in U.S. history is far too ambitious a statement to make.

A better version might look like this:

> The civil rights movement has greatly increased the career opportunities available for Black Americans.

The unnecessary language and details are removed, and the claim can now be measured and supported.

Support

Once the main idea or proposition is stated, the writer attempts to prove or *support* the claim with text evidence and supporting details.

Take for example the sentence, "Seat belts save lives." Though most people can't argue with this statement, its impact on the reader is much greater when supported by additional content. The writer can support this idea by:

- Providing statistics on the rate of highway fatalities alongside statistics for estimated seat belt usage.

- Explaining the science behind a car accident and what happens to a passenger who doesn't use a seat belt.

- Offering anecdotal evidence or true stories from reliable sources on how seat belts prevent fatal injuries in car crashes.

However, using only one form of supporting evidence is not nearly as effective as using a variety to support a claim. Presenting only a list of statistics can be boring to the reader, but providing a true story that's both interesting and humanizing helps. In addition, one example isn't always enough to prove the writer's larger point, so combining it with other examples is extremely effective for the writing. Thus, when reading a passage, don't just look for a single form of supporting evidence.

Another key aspect of supporting evidence is a *reliable source*. Does the writer include the source of the information? If so, is the source well known and trustworthy? Is there a potential for bias? For example, a seat belt study done by a seat belt manufacturer may have its own agenda to promote.

Effective Language Use

Language can be analyzed in a variety of ways. But one of the primary ways is its effectiveness in communicating and especially convincing others.

Rhetoric is a literary technique used to make the writing (or speaking) more effective or persuasive. Rhetoric makes use of other effective language devices such as irony, metaphors, allusion, and repetition. An example of the rhetorical use of repetition would be: "Let go, I say, let go!!!".

Figures of Speech

A *figure of speech* (sometimes called an *idiom*) is a rhetorical device. It's a phrase that's not intended to be taken literally.

When the writer uses a figure of speech, their intention must be clear if it's to be used effectively. Some phrases can be interpreted in a number of ways, causing confusion for the reader. In the PSAT Writing and Language Test, questions may ask for an alternative to a problematic word or phrase. Look for clues to the writer's true intention to determine the best replacement. Likewise, some figures of speech may seem out of place in a more formal piece of writing. To show this, here is the previous seat belt example but with one slight change:

Seat belts save more lives than any other automobile safety feature. Many studies show that airbags save lives as well. However, not all cars have airbags. For instance, some older cars don't. In addition, air bags aren't entirely reliable. For example, studies show that in 15% of accidents, airbags don't deploy as designed. But, on the other hand, seat belt malfunctions happen once in a blue moon.

Most people know that "once in a blue moon" refers to something that rarely happens. However, because the rest of the paragraph is straightforward and direct, using this figurative phrase distracts the reader. In this example, the earlier version is much more effective.

Now it's important to take a moment and review the meaning of the word *literally*. This is because it's one of the most misunderstood and misused words in the English language. *Literally* means that something is exactly what it says it is, and there can be no interpretation or exaggeration. Unfortunately, *literally* is often used for emphasis as in the following example:

This morning, I literally couldn't get out of bed.

This sentence meant to say that the person was extremely tired and wasn't able to get up. However, the sentence can't *literally* be true unless that person was tied down to the bed, paralyzed, or affected by a strange situation that the writer (most likely) didn't intend. Here's another example:

I literally died laughing.

The writer tried to say that something was very funny. However, unless they're writing this from beyond the grave, it can't *literally* be true.

Rhetorical Fallacies

A *rhetorical fallacy* is an argument that doesn't make sense. It usually involves distracting the reader from the issue at hand in some way. There are many kinds of rhetorical fallacies. Here are just a few, along with examples of each:

- *Ad Hominem*: Makes an irrelevant attack against the person making the claim, rather than addressing the claim itself.

- Senator Wilson opposed the new seat belt legislation, but should we really listen to someone who's been divorced four times?

- *Exaggeration*: Represents an idea or person in an obviously excessive manner.

- Senator Wilson opposed the new seat belt legislation. Maybe she thinks if more people die in car accidents, it will help with overpopulation.

- *Stereotyping (or Categorical Claim)*: Claims that all people of a certain group are the same in some way.

- Senator Wilson still opposes the new seat belt legislation. You know women can never admit when they're wrong.

When examining a possible rhetorical fallacy, carefully consider the point the writer is trying to make and if the argument directly relates to that point. If something feels wrong, there's a good chance that a fallacy is at play. The PSAT Writing and Language Test doesn't expect the fallacy to be named using specific terms like those above. However, questions can include identifying why something is a fallacy or suggesting a sounder argument.

Style, Tone, and Mood

Style, *tone*, and *mood* are often thought to be the same thing. Though they're closely related, there are important differences to keep in mind. The easiest way to do this is to remember that style "creates and affects" tone and mood. More specifically, style is *how the writer uses words* to create the desired tone and mood for their writing.

Style

Style can include any number of technical writing choices, and some may have to be analyzed on the test. A few examples of style choices include:

- Sentence Construction: When presenting facts, does the writer use shorter sentences to create a quicker sense of the supporting evidence, or do they use longer sentences to elaborate and explain the information?

- Technical Language: Does the writer use jargon to demonstrate their expertise in the subject, or do they use ordinary language to help the reader understand things in simple terms?

- Formal Language: Does the writer refrain from using contractions such as *won't* or *can't* to create a more formal tone, or do they use a colloquial, conversational style to connect to the reader?

- Formatting: Does the writer use a series of shorter paragraphs to help the reader follow a line of argument, or do they use longer paragraphs to examine an issue in great detail and demonstrate their knowledge of the topic?

On the test, examine the writer's style and how their writing choices affect the way the passage comes across.

Tone

Tone refers to the writer's attitude toward the subject matter. Tone is usually explained in terms of a work of fiction. For example, the tone conveys how the writer feels about their characters and the situations in which they're involved. Nonfiction writing is sometimes thought to have no tone at all, but this is incorrect.

A lot of nonfiction writing has a neutral tone, which is an extremely important tone for the writer to take. A neutral tone demonstrates that the writer is presenting a topic impartially and letting the information speak for itself. On the other hand, nonfiction writing can be just as effective and appropriate if the tone isn't neutral. For instance, take the previous examples involving seat belt use. In them, the writer mostly chooses to retain a neutral tone when presenting information. If the writer would instead include their own personal experience of losing a friend or family member in a car accident, the tone would change dramatically. The tone would no longer be neutral. Now it would show that the writer has a personal stake in the content, allowing them to interpret the information in a different way. When analyzing tone, consider what the writer is trying to achieve in the passage, and how they *create* the tone using style.

Mood

Mood refers to the feelings and atmosphere that the writer's words create for the reader. Like tone, many nonfiction pieces can have a neutral mood. To return to the previous example, if the writer would

45

choose to include information about a person they know being killed in a car accident, the passage would suddenly carry an emotional component that is absent in the previous examples. Depending on how they present the information, the writer can create a sad, angry, or even hopeful mood. When analyzing the mood, consider what the writer wants to accomplish and whether the best choice was made to achieve that end.

Consistency

Whatever style, tone, and mood the writer uses, good writing should remain *consistent* throughout. If the writer chooses to include the tragic, personal experience above, it would affect the style, tone, and mood of the entire piece. It would seem out of place for such an example to be used in the middle of a neutral, measured, and analytical piece. To adjust the rest of the piece, the writer needs to make additional choices to remain consistent. For example, the writer might decide to use the word *tragedy* in place of the more neutral *fatality*, or they could describe a series of car-related deaths as an *epidemic*. Adverbs and adjectives such as *devastating* or *horribly* could be included to maintain this consistent attitude toward the content. When analyzing writing, look for sudden shifts in style, tone, and mood, and consider whether the writer would be wiser to maintain the prevailing strategy.

Syntax

Syntax is the order of words in a sentence. While most of the writing on the test has proper syntax, there may be questions on ways to vary the syntax for effectiveness. One of the easiest writing mistakes to spot is *repetitive sentence structure*. For example:

> Seat belts are important. They save lives. People don't like to use them. We have to pass seat belt laws. Then more people will wear seat belts. More lives will be saved.

What's the first thing that comes to mind when reading this example? The short, choppy, and repetitive sentences! In fact, most people notice this syntax issue more than the content itself. By combining some sentences and changing the syntax of others, the writer can create a more effective writing passage:

> Seat belts are important because they save lives. Since people don't like to use seat belts, though, more laws requiring their usage need to be passed. Only then will more people wear them and only then will more lives be saved.

Many rhetorical devices can be used to vary syntax (more than can possibly be named here). These often have intimidating names like *anadiplosis*, *metastasis*, and *paremptosis*. The test questions don't ask for definitions of these tricky techniques, but they can ask how the writer plays with the words and what effect that has on the writing. For example, *anadiplosis* is when the last word (or phrase) from a sentence is used to begin the next sentence:

> Cars are driven by people. People cause accidents. Accidents cost taxpayers money.

The test doesn't ask for this technique by name, but be prepared to recognize what the writer is doing and why they're using the technique in this situation. In this example, the writer is probably using *anadiplosis* to demonstrate causation.

Quantitative Information

Some writing in the test contains *infographics* such as charts, tables, or graphs. In these cases, interpret the information presented and determine how well it supports the claims made in the text. For example,

if the writer makes a case that seat belts save more lives than other automobile safety measures, they might want to include a graph (like the one below) showing the number of lives saved by seat belts versus those saved by air bags.

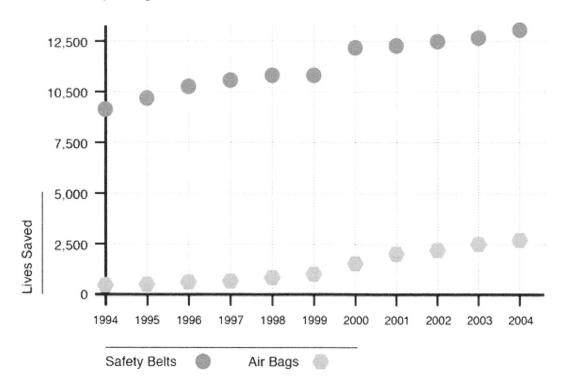

Based on data from the National Highway Traffic Safety Administration

If the graph clearly shows a higher number of lives are saved by seat belts, then it's effective. However, if the graph shows air bags save more lives than seat belts, then it doesn't support the writer's case.

Finally, graphs should be easy to understand. Their information should immediately be clear to the reader at a glance. Here are some basic things to keep in mind when interpreting infographics:

- In a *bar graph*, higher bars represent larger numbers. Lower bars represent smaller numbers.

- *Line graphs* are the same, but often show trends over time. A line that consistently ascends from left to right shows a steady increase over time. A line that consistently descends from left to right shows a steady decrease over time. If the line bounces up and down, this represents instability or inconsistency in the trend. When interpreting a line graph, determine the point the writer is trying to make, and then see if the graph supports that point.

- *Pie charts* are used to show proportions or percentages of a whole, but are less effective in showing change over time.

- *Tables* present information in numerical form, not as graphics. When interpreting a table, make sure to look for patterns in the numbers.

There can also be timelines, illustrations, or maps on the test. When interpreting these, keep in mind the writer's intentions and determine whether or not the graphic supports the case.

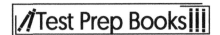

Standard English Conventions

Most of the topics discussed so far deal with the writer's choices and their effectiveness in a particular writing piece. In many cases, even ineffective writing can be grammatically correct. The following sections examine writing problems that actually break the rules of Standard English. These aren't questions of intent or judgment calls by the writer. These are mistakes that *must* be corrected.

Sentence Structure

Fragments and Run-Ons

A *sentence fragment* is a failed attempt to create a complete sentence because it's missing a required noun or verb. Fragments don't function properly because there isn't enough information to understand the writer's intended meaning. For example:

> Seat belt use corresponds to a lower rate of hospital visits, reducing strain on an already overburdened healthcare system. Insurance claims as well.

Look at the last sentence: *Insurance claims as well*. What does this mean? This is a fragment because it has a noun but no verb, and it leaves the reader guessing what the writer means about insurance claims. Many readers can probably infer what the writer means, but this distracts them from the flow of the writer's argument. Choosing a suitable replacement for a sentence fragment may be one of the questions on the test. The fragment is probably related to the surrounding content, so look at the overall point the writer is trying to make and choose the answer that best fits that idea.

Remember that sometimes a fragment can *look* like a complete sentence or have all the nouns and verbs it needs to make sense. Consider the following two examples:

> Seat belt use corresponds to a lower rate of hospital visits.

> Although seat belt use corresponds to a lower rate of hospital visits.

Both examples above have nouns and verbs, but only the first sentence is correct. The second sentence is a fragment, even though it's actually longer. The key is the writer's use of the word *although*. Starting a sentence with *although* turns that part into a *subordinate clause* (more on that next). Keep in mind that one doesn't have to remember that it's called a subordinate clause on the test. Just be able to recognize that the words form an incomplete thought and identify the problem as a sentence fragment.

A *run-on sentence* is, in some ways, the opposite of a fragment. It contains two or more sentences that have been improperly forced together into one. An example of a run-on sentence looks something like this:

> Seat belt use corresponds to a lower rate of hospital visits it also leads to fewer insurance claims.

Here, there are two separate ideas in one sentence. It's difficult for the reader to follow the writer's thinking because there is no transition from one idea to the next. On the test, choose the best way to correct the run-on sentence.

Here are two possibilities for the sentence above:

- Seat belt use corresponds to a lower rate of hospital visits. It also leads to fewer insurance claims.

- Seat belt use corresponds to a lower rate of hospital visits, but it also leads to fewer insurance claims.

Both solutions are grammatically correct, so which one is the best choice? That depends on the point that the writer is trying to make. Always read the surrounding text to determine what the writer wants to demonstrate, and choose the option that best supports that thought.

Subordination and Coordination

With terms like "coordinate clause" and "subordinating conjunction," grammar terminology can scare people! So, just for a minute, forget about the terms and look at how the sentences work.

Sometimes a sentence has two ideas that work together. For example, say the writer wants to make the following points:

Seat belt laws have saved an estimated 50,000 lives.

More lives are saved by seat belts every year.

These two ideas are directly related and appear to be of equal importance. Therefore they can be joined with a simple "and" as follows:

Seat belt laws have saved an estimated 50,000 lives, and more lives are saved by seat belts every year.

The word *and* in the sentence helps the two ideas work together or, in other words, it "coordinates" them. It also serves as a junction where the two ideas come together, better known as a *conjunction*. Therefore the word *and* is known as a *coordinating conjunction* (a word that helps bring two equal ideas together). Now that the ideas are joined together by a conjunction, they are known as *clauses*. Other coordinating conjunctions include *or*, *but*, and *so*.

Sometimes, however, two ideas in a sentence are *not* of equal importance:

Seat belt laws have saved an estimated 50,000 lives.

Many more lives could be saved with stronger federal seat belt laws.

In this case, combining the two with a coordinating conjunction (*and*) creates an awkward sentence:

Seat belt laws have saved an estimated 50,000 lives, and many more lives could be saved with stronger federal seat belt laws.

Now the writer uses a word to show the reader which clause is the most important (or the "boss") of the sentence:

Although seat belt laws have saved an estimated 50,000 lives, many more lives could be saved with stronger federal seat belt laws.

49

In this example, the second clause is the key point that the writer wants to make, and the first clause works to set up that point. Since the first clause "works for" the second, it's called the *subordinate clause*. The word *although* tells the reader that this idea isn't as important as the clause that follows. This word is called the *subordinating conjunction*. Other subordinating conjunctions include *after*, *because, if, since, unless*, and many more. As mentioned before, it's easy to spot subordinate clauses because they don't stand on their own (as shown in this previous example):

> Although seat belt laws have saved an estimated 50,000 lives

This is not a complete thought. It needs the other clause (called the *independent clause*) to make sense. On the test, when asked to choose the best subordinating conjunction for a sentence, look at the surrounding text. Choose the word that best allows the sentence to support the writer's argument.

Parallel Structure

Parallel structure usually has to do with lists. Look at the following sentence and spot the mistake:

> Increased seat belt legislation has been supported by the automotive industry, the insurance industry, and doctors.

Many people don't see anything wrong, but the word *doctors* breaks the sentence's parallel structure. The previous items in the list refer to an industry as a singular noun, so every item in the list must follow that same format:

> Increased seat belt legislation has been supported by the automotive industry, the insurance industry, and the healthcare industry.

Another common mistake in parallel structure might look like this:

> Before the accident, Maria enjoyed swimming, running, and played soccer.

Here, the words "played soccer" break the parallel structure. To correct it, the writer must change the final item in the list to match the format of the previous two:

> Before the accident, Maria enjoyed swimming, running, and playing soccer.

Usage

Modifier Placement

Modifiers are words or phrases (often adjectives or nouns) that add detail to, explain, or limit the meaning of other parts of a sentence. Look at the following example:

> A big pine tree is in the yard.

In the sentence, the words *big* (an adjective) and *pine* (a noun) modify *tree* (the head noun).

All related parts of a sentence must be placed together correctly. *Misplaced* and *dangling modifiers* are common writing mistakes. In fact, they're so common that many people are accustomed to seeing them and can decipher an incorrect sentence without much difficulty. On the test, expect to be asked to identify and correct this kind of error.

Misplaced Modifiers

Since *modifiers* refer to something else in the sentence (*big* and *pine* refer to *tree* in the example above), they need to be placed close to what they modify. If a modifier is so far away that the reader isn't sure what it's describing, it becomes a *misplaced modifier*. For example:

> Seat belts almost saved 5,000 lives in 2009.

It's likely that the writer means that the total number of lives saved by seat belts in 2009 is close to 5,000. However, due to the misplaced modifier (*almost*), the sentence actually says there are 5,000 instances when seat belts *almost saved lives*. In this case, the position of the modifier is actually the difference between life and death (at least in the meaning of the sentence). A clearer way to write the sentence is:

> Seat belts saved almost 5,000 lives in 2009.

Now that the modifier is close to the 5,000 lives it references, the sentence's meaning is clearer.

Another common example of a misplaced modifier occurs when the writer uses the modifier to begin a sentence. For example:

> Having saved 5,000 lives in 2009, Senator Wilson praised the seat belt legislation.

It seems unlikely that Senator Wilson saved 5,000 lives on her own, but that's what the writer is saying in this sentence. To correct this error, the writer should move the modifier closer to the intended object it modifies. Here are two possible solutions:

> Having saved 5,000 lives in 2009, the seat belt legislation was praised by Senator Wilson.

> Senator Wilson praised the seat belt legislation, which saved 5,000 lives in 2009.

When choosing a solution for a misplaced modifier, look for an option that places the modifier close to the object or idea it describes.

Dangling Modifiers

A modifier must have a target word or phrase that it's modifying. Without this, it's a *dangling modifier*. Dangling modifiers are usually found at the beginning of sentences:

> After passing the new law, there is sure to be an improvement in highway safety.

This sentence doesn't say anything about who is passing the law. Therefore, "After passing the new law" is a dangling modifier because it doesn't modify anything in the sentence. To correct this type of error, determine what the writer intended the modifier to point to:

> After passing the new law, legislators are sure to see an improvement in highway safety.

"After passing the new law" now points to *legislators*, which makes the sentence clearer and eliminates the dangling modifier.

Shifts in Construction

It's been said several times already that *good writing must be consistent*. Another common writing mistake occurs when the writer unintentionally shifts verb tense, voice, or noun-pronoun agreement. This shift can take place within a sentence, within a paragraph, or over the course of an entire piece of writing. On the test, questions may ask that this kind of error be identified. Here are some examples.

Shift in Verb Tense

Even though test questions don't ask for verb tenses to be identified, they may cover recognizing when these tenses change unexpectedly:

During the accident, the airbags malfunction and the passengers were injured.

In this sentence, the writer unintentionally shifts from present tense ("airbags malfunction" is happening *now)* to past tense ("passengers were injured" has *already happened*.) This is very confusing. To correct this error, the writer must stay in the same tense throughout. Two possible solutions are:

During the accident, the airbags malfunctioned and the passengers were injured.

During the accident, the airbags malfunction and the passengers are injured.

Shift in Voice

Sometimes the writer accidentally slips from active voice to passive voice in the middle of a sentence. This is a difficult mistake to catch because it's something people often do when speaking to one another. First, it's important to understand the difference between active and passive voice. Most sentences are written in *active voice*, which means that the noun is doing what the verb in the sentence says. For example:

Seat belts save lives.

Here, the noun (*seat belt*) is doing the saving. However, in *passive voice*, the verb is doing something to the noun:

Lives are saved.

In this case, the noun (*lives*) is the thing *being saved*. Passive voice is difficult for many people to identify and understand, but there's a simple (and memorable) way to check: simply add "by zombies" to the end of the verb and, if it makes sense, then the verb is written in passive voice. For example: "My car was wrecked...by zombies." Also, in the above example, "Lives are saved...by zombies." If the zombie trick doesn't work, then the sentence is in active voice.

Here's what a shift in voice looks like in a sentence:

When Amy buckled her seat belt, a satisfying click was heard.

The writer shifts from active voice in the beginning of the sentence to passive voice after the comma (remember, "a satisfying click was heard...by zombies"). To fix this mistake, the writer must remain in active voice throughout:

When Amy buckled her seat belt, she heard a satisfying click.

This sentence is now grammatically correct, easier to read...and zombie free!

Shift in Noun-Pronoun Agreement

Pronouns are used to replace nouns so sentences don't have a lot of unnecessary repetition. This repetition can make a sentence seem awkward as in the following example:

> Seat belts are important because seat belts save lives, but seat belts can't do so unless seat belts are used.

Replacing some of the nouns (*seat belts*) with a pronoun (*they*) improves the flow of the sentence:

> Seat belts are important because they save lives, but they can't do so unless they are used.

A pronoun should agree in number (singular or plural) with the noun that precedes it. Another common writing error is the shift in *noun-pronoun agreement*. Here's an example:

> When people are getting in a car, he should always remember to buckle his seatbelt.

The first half of the sentence talks about a plural (*people*), while the second half refers to a singular person (*he* and *his*). These don't agree, so the sentence should be rewritten as:

> When people are getting in a car, they should always remember to buckle their seatbelt.

Pronouns

Pronoun Person

Pronoun person refers to the narrative voice the writer uses in a piece of writing. A great deal of nonfiction is written in third person, which uses pronouns like *he, she, it,* and *they* to convey meaning. Occasionally a writer uses first person (*I, me, we,* etc.) or second person (*you*). Any choice of pronoun person can be appropriate for a particular situation, but the writer must remain consistent and logical.

Test questions may cover examining samples that should stay in a single pronoun person, be it first, second, or third. Look out for shifts between words like *you* and *I* or *he* and *they*.

Pronoun Clarity

Pronouns always refer back to a noun. However, as the writer composes longer, more complicated sentences, the reader may be unsure which noun the pronoun should replace. For example:

> An amendment was made to the bill, but now it has been voted down.

Was the amendment voted down or the entire bill? It's impossible to tell from this sentence. To correct this error, the writer needs to restate the appropriate noun rather than using a pronoun:

> An amendment was made to the bill, but now the bill has been voted down.

Pronouns in Combination

Writers often make mistakes when choosing pronouns to use in combination with other nouns. The most common mistakes are found in sentences like this:

> Please join Senator Wilson and I at the event tomorrow.

Notice anything wrong? Though many people think the sentence sounds perfectly fine, the use of the pronoun *I* is actually incorrect. To double-check this, take the other person out of the sentence:

Please join I at the event tomorrow.

Now the sentence is obviously incorrect, as it should read, "Please join *me* at the event tomorrow." Thus, the first sentence should replace *I* with *me*:

Please join Senator Wilson and me at the event tomorrow.

For many people, this sounds wrong because they're used to hearing and saying it incorrectly. Take extra care when answering this kind of question and follow the double-checking procedure.

Agreement

In English writing, certain words connect to other words. People often learn these connections (or *agreements*) as young children and use the correct combinations without a second thought. However, the questions on the test dealing with agreement probably aren't simple ones.

Subject-Verb Agreement

Which of the following sentences is correct?

A large crowd of protesters was on hand.

A large crowd of protesters were on hand.

Many people would say the second sentence is correct, but they'd be wrong. However, they probably wouldn't be alone. Most people just look at two words: *protesters were*. Together they make sense. They sound right. The problem is that the verb *were* doesn't refer to the word *protesters*. Here, the word *protesters* is part of a prepositional phrase that clarifies the actual subject of the sentence (*crowd*). Take the phrase "of protesters" away and re-examine the sentences:

A large crowd was on hand.

A large crowd were on hand.

Without the prepositional phrase to separate the subject and verb, the answer is obvious. The first sentence is correct. On the test, look for confusing prepositional phrases when answering questions about subject-verb agreement. Take the phrase away, and then recheck the sentence.

Noun Agreement

Nouns that refer to other nouns must also match in number. Take the following example:

John and Emily both served as an intern for Senator Wilson.

Two people are involved in this sentence: John and Emily. Therefore, the word *intern* should be plural to match. Here is how the sentence should read:

John and Emily both served as interns for Senator Wilson.

Frequently Confused Words

There are a handful of words in the English language that writers often confuse with other words because they sound similar or identical. Errors involving these words are hard to spot because they *sound* right even when they're wrong. Also, because these mistakes are so pervasive, many people think they're correct. Here are a few examples that may be encountered on the test:

They're vs. Their vs. There

This set of words is probably the all-time winner of misuse. The word *they're* is a contraction of "they are." Remember that contractions combine two words, using an apostrophe to replace any eliminated letters. If a question asks whether the writer is using the word *they're* correctly, change the word to "they are" and reread the sentence. Look at the following example:

> Legislators can be proud of they're work on this issue.

This sentence *sounds* correct, but replace the contraction *they're* with "they are" to see what happens:

> Legislators can be proud of they are work on this issue.

The result doesn't make sense, which shows that it's an incorrect use of the word *they're*. Did the writer mean to use the word *their* instead? The word *their* indicates possession because it shows that something *belongs* to something else. Now put the word *their* into the sentence:

> Legislators can be proud of their work on this issue.

To check the answer, find the word that comes right after the word *their* (which in this case is *work*). Pose this question: whose *work* is it? If the question can be answered in the sentence, then the word signifies possession. In the sentence above, it's the legislators' work. Therefore, the writer is using the word *their* correctly.

If the words *they're* and *their* don't make sense in the sentence, then the correct word is almost always *there*. The word *there* can be used in many different ways, so it's easy to remember to use it when *they're* and *their* don't work. Now test these methods with the following sentences:

> Their going to have a hard time passing these laws.

> Enforcement officials will have there hands full.

> They're are many issues to consider when discussing car safety.

In the first sentence, asking the question "Whose going is it?" doesn't make sense. Thus the word *their* is wrong. However, when replaced with the conjunction *they're* (or *they are*), the sentence works. Thus the correct word for the first sentence should be *they're*.

In the second sentence, ask this question: "Whose hands are full?" The answer (*enforcement officials*) is correct in the sentence. Therefore, the word *their* should replace *there* in this sentence.

In the third sentence, changing the word *they're* to "they are" ("They are are many issues") doesn't make sense. Ask this question: "Whose are is it?" This makes even less sense, since neither of the words *they're* or *their* makes sense. Therefore, the correct word must be *there*.

Who's vs. Whose

Who's is a contraction of "who is" while the word *whose* indicates possession. Look at the following sentence:

>Who's job is it to protect America's drivers?

The easiest way to check for correct usage is to replace the word *who's* with "who is" and see if the sentence makes sense:

>Who is job is it to protect America's drivers?

By changing the contraction to "Who is" the sentence no longer makes sense. Therefore, the correct word must be *whose*.

Your vs. You're

The word *your* indicates possession, while *you're* is a contraction for "you are." Look at the following example:

>Your going to have to write your congressman if you want to see action.

Again, the easiest way to check correct usage is to replace the word *Your* with "You are" and see if the sentence still makes sense.

>You are going to have to write your congressman if you want to see action.

By replacing Your with "You are," the sentence still makes sense. Thus, in this case, the writer should have used "You're."

Its vs. It's

Its is a word that indicates possession, while the word *it's* is a contraction of "it is." Once again, the easiest way to check for correct usage is to replace the word with "it is" and see if the sentence makes sense. Look at the following sentence:

>It's going to take a lot of work to pass this law.

Replacing *it's* with "it is" results in this: "It is going to take a lot of work to pass this law." This makes sense, so the contraction (*it's*) is correct. Now look at another example:

>The car company will have to redesign it's vehicles.

Replacing *it's* with "it is" results in this: "The car company will have to redesign it is vehicles." This sentence doesn't make sense, so the contraction (*it's*) is incorrect.

Than vs. Then

Than is used in sentences that involve comparisons, while *then* is used to indicate an order of events. Consider the following sentence:

>Japan has more traffic fatalities than the U.S.

The use of the word *than* is correct because it compares Japan to the U.S. Now look at another example:

>Laws must be passed, and then we'll see a change in behavior.

Here the use of the word *then* is correct because one thing happens after the other.

Affect vs. Effect

Affect is a verb that means to change something, while *effect* is a noun that indicates such a change. Look at the following sentence:

> There are thousands of people affected by the new law.

This sentence is correct because *affected* is a verb that tells what's happening. Now look at this sentence:

> The law will have a dramatic effect.

This sentence is also correct because *effect* is a noun and the thing that happens.

Note that a noun version of *affect* is occasionally used. It means "emotion" or "desire," usually in a psychological sense.

Two vs. Too vs. To

Two is the number (2). *Too* refers to an amount of something, or it can mean *also*. *To* is used for everything else. Look at the following sentence:

> Two senators still haven't signed the bill.

This is correct because there are *two* (2) senators. Here's another example:

> There are too many questions about this issue.

In this sentence, the word *too* refers to an amount ("too many questions"). Now here's another example:

> Senator Wilson is supporting this legislation, too.

In this sentence, the word *also* can be substituted for the word *too*, so it's also correct. Finally, one last example:

> I look forward to signing this bill into law.

In this sentence, the tests for *two* and *too* don't work. Thus the word *to* fits the bill!

Other Common Writing Confusions

In addition to all of the above, there are other words that writers often misuse. This doesn't happen because the words sound alike, but because the writer is not aware of the proper way to use them.

Logical Comparison

Writers often make comparisons in their writing. However, it's easy to make mistakes in sentences that involve comparisons, and those mistakes are difficult to spot. Try to find the error in the following sentence:

> Senator Wilson's proposed seat belt legislation was similar to Senator Abernathy.

Can't find it? First, ask what two things are actually being compared. It seems like the writer *wants* to compare two different types of legislation, but the sentence actually compares legislation ("Senator Wilson's proposed seat belt legislation") to a person ("Senator Abernathy"). This is a strange and illogical comparison to make.

So how can the writer correct this mistake? The answer is to make sure that the second half of the sentence logically refers back to the first half. The most obvious way to do this is to repeat words:

> Senator Wilson's proposed seat belt legislation was similar to Senator Abernathy's seat belt legislation.

Now the sentence is logically correct, but it's a little wordy and awkward. A better solution is to eliminate the word-for-word repetition by using suitable replacement words:

> Senator Wilson's proposed seat belt legislation was similar to that of Senator Abernathy.

> Senator Wilson's proposed seat belt legislation was similar to the bill offered by Senator Abernathy.

Here's another similar example:

> More lives in the U.S. are saved by seat belts than Japan.

The writer probably means to compare lives saved by seat belts in the U.S. to lives saved by seat belts in Japan. Unfortunately, the sentence's meaning is garbled by an illogical comparison, and instead refers to U.S. lives saved *by Japan* rather than *in Japan.* To resolve this issue, first repeat the words and phrases needed to make an identical comparison:

> More lives in the U.S. are saved by seat belts than lives in Japan are saved by seat belts.

Then, use a replacement word to clean up the repetitive text:

> More lives in the U.S. are saved by seat belts than in Japan.

Punctuation

On the test there may be a sentence where all the words are correct, but the writer uses *punctuation* incorrectly. It probably won't be something as simple as a missing period or question mark. Instead it could be one of the commonly misunderstood punctuation marks.

Colons

Colons can be used in the following situations and examples:

- To introduce lists
- Carmakers have three choices: improve seat belt design, pay financial penalties, or go out of business.
- To introduce new ideas
- There is only one person who can champion this legislation: Senator Wilson.
- To separate titles and subtitles
- Show Some Restraint: The History of Seat Belts

Semicolons

Semicolons can be used in the following situations:

- To separate two related independent clauses
- The proposed bill was voted down; opponents were concerned about the tax implications.

  Note: These are known as *independent clauses* because each one stands on its own as a complete sentence. Semicolons *cannot* be used to separate an independent clause from a dependent clause, nor to separate two dependent clauses.

- To separate complex items in a list
- Joining Senator Wilson onstage were Jim Robinson, head of the NHTSA; Kristin Gabber, a consumer advocate; and Milton Webster, an accident survivor.

  Note that while items in a list are usually separated by commas, readers can easily get confused if the list items themselves contain internal commas.

Hyphens vs. Dashes

Hyphens (-) and *dashes* (–) are not the same. *Hyphens* are shorter, and they help combine or clarify words in certain situations like:

- Creating an adjective: *safety-conscious*
- Creating compound numbers: *fifty-nine*
- Avoiding confusion with another word: *re-sent* vs. *resent*
- Avoiding awkward letter combinations: *semi-intellectual* vs. *semiintellectual*

Dashes are longer and show an interruption in the flow of the sentence. In this context, they can be used much the same way as commas or parentheses:

- The legislation—which was supported by 80 percent of Americans—did not pass.

Commas

Commas are used in many different situations. Here are some of the most misunderstood examples:

- Separating simple items in a list
- The legislation had the support of Republicans, Democrats, and Independents.
- Separating adjectives that modify the same noun
- The weak, meaningless platitudes had no effect on the listeners.

- Separating independent and dependent clauses
- After passing the bill, the lawmakers celebrated.
- Note: "After passing the bill" is a *dependent clause* because it's not a complete sentence on its own.
- Separating quotations from introductory text
- Senator Wilson asked, "How can we get this bill passed?"
- Showing interruption in the flow of a sentence. In this context, commas can be used in the same way as semicolons or parentheses.
- The legislation, which was supported by 80 percent of Americans, did not pass.
- Note: Commas cannot be used if the clause or phrase in question is essential to the meaning of the sentence.

During the test, it may be hard to remember all the rules for comma usage. Read the sentence and listen to its ebb and flow. If a particular answer looks, sounds, or feels wrong for some reason, there's probably a good reason for it. Look at another option instead.

Apostrophes

Apostrophes are often misused. For the purpose of the test, there are three things to know about using apostrophes:

- Use apostrophes to show possession
- Senator Wilson's bill just passed committee.
- Use apostrophes in contractions to replace eliminated letters
- Does not → Doesn't

Note: It's common to see acronyms made plural using apostrophes (RV's, DVD's, TV's), but these are incorrect. Acronyms function as words, so they are pluralized the same way (RVs, DVDs, TVs).

On the test, when an apostrophe-related question is asked, determine if it shows possession or is part of a contraction. If neither answer fits, then the apostrophe probably doesn't belong there.

Final Tips

Usage Conventions

On the test, don't overlook simple, obvious writing errors such as these:

- Is the first word in a sentence capitalized?
- Are countries, geographical features, and proper nouns capitalized?
- Conversely, are words capitalized that should *not* be?
- Do sentences end with proper punctuation marks?
- Are commas and quotation marks used appropriately?
- Do contractions include apostrophes?
- Are apostrophes used for plurals? (Almost never!)

Look for Context

Keep in mind that the test may give several choices to replace a writing selection, and all of them may be grammatically correct. In such cases, choose the answer that makes the most sense in the context of

the piece. What's the writer trying to say? What's their main idea? Look for the answer that best supports this theme.

Use Your Instincts

With the few notable exceptions above, instinct is often the best guide to spotting writing problems. If something sounds wrong, then it may very well be wrong. The good thing about a test like this is that the problem doesn't have to be labeled as an example of "faulty parallelism" or "improper noun-pronoun agreement." It's enough just to recognize that a problem exists and choose the best solution.

Take a Break

After reading and thinking about all of these aspects of grammar so intensely, the brain may start shutting down. If the words aren't making sense, or reading the same sentence several times still has no meaning, it's time to stop. Take a thirty-second vacation. Forget about grammar, syntax, and writing for half a minute to clear the mind. Take a few deep breaths and think about something to do after the test is over. It's surprising how quickly the brain refreshes itself!

Practice Questions

Questions 1-5 are based on the following passage:

(1) <u>One of the icon's of romantic and science fiction literature</u> remains Mary Shelley's classic, *Frankenstein, or The Modern Prometheus*. Schools throughout the world still teach the book in literature and philosophy courses. Scientific communities also engage in discussion on the novel. But why? Besides the novel's engaging writing style the story's central theme remains highly relevant in a world of constant discovery and moral dilemmas. Central to the core narrative is the (2) <u>struggle between enlightenment and the cost of overusing power.</u>

The subtitle, *The Modern Prometheus*, encapsulates the inner theme of the story more than the main title of *Frankenstein*. As with many romantic writers, Shelley invokes the classical myths and symbolism of Ancient Greece and Rome to high light core ideas. Looking deeper into the myth of Prometheus sheds light not only on the character of Frankenstein (3) <u>but also poses a psychological dilemma to the audience.</u> Prometheus is the titan who gave fire to mankind. However, more than just fire he gave people knowledge and power. The power of fire advanced civilization. Yet, for giving fire to man, Prometheus is (4) <u>punished by the gods bound to a rock and tormented for his act</u>. This is clearly a parallel to Frankenstein—he is the modern Prometheus.

Frankenstein's quest for knowledge becomes an obsession. It leads him to literally create new life, breaking the bounds of conceivable science to illustrate that man can create life out of nothing. Yet he ultimately faltered as a creator, abandoning his progeny in horror of what he created. Frankenstein then suffers his creature's wrath, (5) <u>the result of his pride, obsession for power and lack of responsibility.</u>

Shelley isn't condemning scientific achievement. Rather, her writing reflects that science and discovery are good things, but, like all power, it must be used wisely. The text alludes to the message that one must have reverence for nature and be mindful of the potential consequences. Frankenstein did not take responsibility or even consider how his actions would affect others. His scientific brilliance ultimately led to suffering.

1. Which of the following would be the best choice for this sentence (reproduced below)?

> (1) <u>One of the icon's of romantic and science fiction literature</u> remains Mary Shelley's classic, Frankenstein, or The Modern Prometheus.
>
> a. NO CHANGE
> b. One of the icons of romantic and science fiction literature
> c. One of the icon's of romantic, and science fiction literature,
> d. The icon of romantic and science fiction literature

2. Which of the following would be the best choice for this sentence (reproduced below)?

Central to the core narrative is the (2) struggle between enlightenment and the cost of overusing power.

a. NO CHANGE
b. struggle between enlighten and the cost of overusing power.
c. struggle between enlightenment's cost of overusing power.
d. struggle between enlightening and the cost of overusing power.

3. Which of the following would be the best choice for this sentence (reproduced below)?

Looking deeper into the myth of Prometheus sheds light not only on the character of Frankenstein (3) but also poses a psychological dilemma to the audience.

a. NO CHANGE
b. but also poses a psychological dilemma with the audience.
c. but also poses a psychological dilemma for the audience.
d. but also poses a psychological dilemma there before the audience.

4. Which of the following would be the best choice for this sentence (reproduced below)?

Yet, for giving fire to man, Prometheus is (4) punished by the gods bound to a rock and tormented for his act.

a. NO CHANGE
b. punished by the gods, bound to a rock and tormented for his act.
c. bound to a rock and tormented as punishment by the gods.
d. punished for his act by being bound to a rock and tormented as punishment from the gods.

5. Which of the following would be the best choice for this sentence (reproduced below)?

Frankenstein then suffers his creature's wrath, (5) the result of his pride, obsession for power and lack of responsibility.

a. NO CHANGE
b. the result of his pride, obsession for power and lacking of responsibility.
c. the result of his pride, obsession for power, and lack of responsibility.
d. the result of his pride and also his obsession for power and lack of responsibility.

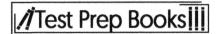

Answer Explanations

1. B: Choice *B* is correct because it removes the apostrophe from *icon's*, since the noun *icon* is not possessing anything. This conveys the author's intent of setting *Frankenstein* apart from other icons of the romantic and science fiction genres. Choices *A* and *C* are therefore incorrect. Choice *D* is a good revision but alters the meaning of the sentence—*Frankenstein* is one of the icons, not the sole icon.

2. A: Choice *A* is correct, as the sentence doesn't require changes. Choice *B* incorrectly changes the noun *enlightenment* into the verb *enlighten*. Choices *C* and *D* alter the original meaning of the sentence.

3. A: Choice *A* is correct, as *not only* and *but also* are correlative pairs. In this sentence, *but* successfully transitions the first part into the second half, making punctuation unnecessary. Additionally, the use of *to* indicates that an idea or challenge is being presented to the reader. Choice *B*'s *with*, *C*'s *for*, and *D*'s *there before* are not as active, meaning these revisions weaken the sentence.

4. C: Choice *C* reverses the order of the section, making the sentence more direct. Choice *A* lacks a comma after *gods*, and although Choice *B* adds this, the structure is too different from the first half of the sentence to flow correctly. Choice *D* is overly complicated and repetitive in its structure even though it doesn't need any punctuation.

5. C: Choice *C* successfully applies a comma after *power*, distinguishing the causes of Frankenstein's suffering and maintaining parallel structure. Choice *A* is thus incorrect. Choice *B* lacks the necessary punctuation and unnecessarily changes *lack* to a gerund. Choice *D* adds unnecessary wording, making the sentence more cumbersome.

Math Test

Heart of Algebra

Creating, Solving, or Interpreting a Linear Expression or Equation in One Variable

Linear expressions and equations are concise mathematical statements that can be written to model a variety of scenarios. Questions found pertaining to this topic will contain one variable only. A variable is an unknown quantity, usually denoted by a letter (x, n, p, etc.). In the case of linear expressions and equations, the power of the variable (its exponent) is 1. A variable without a visible exponent is raised to the first power.

Writing Linear Expressions and Equations

A linear expression is a statement about an unknown quantity expressed in mathematical symbols. The statement "five times a number added to forty" can be expressed as $5x + 40$. A linear equation is a statement in which two expressions (at least one containing a variable) are equal to each other. The statement "five times a number added to forty is equal to ten" can be expressed as $5x + 40 = 10$. Real-world scenarios can also be expressed mathematically. Consider the following:

> Bob had $20 and Tom had $4. After selling 4 ice cream cones to Bob, Tom has as much money as Bob.

The cost of an ice cream cone is an unknown quantity and can be represented by a variable. The amount of money Bob has after his purchase is four times the cost of an ice cream cone subtracted from his original $20. The amount of money Tom has after his sale is four times the cost of an ice cream cone added to his original $4. This can be expressed as: $20 - 4x = 4x + 4$, where x represents the cost of an ice cream cone.

When expressing a verbal or written statement mathematically, it is key to understand words or phrases that can be represented with symbols. The following are examples:

Symbol	Phrase
$+$	added to, increased by, sum of, more than
$-$	decreased by, difference between, less than, take away
x	multiplied by, 3 (4, 5 . . .) times as large, product of
$\div$	divided by, quotient of, half (third, etc.) of
$=$	is, the same as, results in, as much as
$x, t, n, etc.$	a number, unknown quantity, value of

Evaluating and Simplifying Algebraic Expressions

Given an algebraic expression, students may be asked to evaluate for given values of variable(s). In doing so, students will arrive at a numerical value as an answer. For example:

$$\text{Evaluate } a - 2b + ab \ for \ a = 3 \text{ and } b = -1$$

To evaluate an expression, the given values should be substituted for the variables and simplified using the order of operations. In this case: $(3) - 2(-1) + (3)(-1)$. Parentheses are used when substituting.

Given an algebraic expression, students may be asked to simplify the expression. For example:

$$\text{Simplify } 5x^2 - 10x + 2 - 8x^2 + x - 1.$$

Simplifying algebraic expressions requires combining like terms. A term is a number, variable, or product of a number and variables separated by addition and subtraction. The terms in the above expressions are: $5x^2, -10x, 2, -8x^2, x$, and -1. Like terms have the same variables raised to the same powers (exponents). To combine like terms, the coefficients (numerical factor of the term including sign) are added, while the variables and their powers are kept the same. The example above simplifies to $-3x^2 - 9x + 1$.

Solving Linear Equations

When asked to solve a linear equation, it requires determining a numerical value for the unknown variable. Given a linear equation involving addition, subtraction, multiplication, and division, isolation of the variable is done by working backward. Addition and subtraction are inverse operations, as are multiplication and division; therefore, they can be used to cancel each other out.

The first steps to solving linear equations are to distribute if necessary and combine any like terms that are on the same side of the equation. Sides of an equation are separated by an $=$ sign. Next, the equation should be manipulated to get the variable on one side. Whatever is done to one side of an equation, must be done to the other side to remain equal. Then, the variable should be isolated by using inverse operations to undo the order of operations backward. Undo addition and subtraction, then undo multiplication and division. For example:

$$\text{Solve } 4(t - 2) + 2t - 4 = 2(9 - 2t)$$

Steps →

1. Distribute: $4t - 8 + 2t - 4 = 18 - 4t$

2. Combine like terms: $6t - 12 = 18 - 4t$

3. Add 4t to each side to move the variable: $10t - 12 = 18$

4. Add 12 to each side to isolate the variable: $10t = 30$

5. Divide each side by 10 to isolate the variable: $t = 3$

The answer can be checked by substituting the value for the variable into the original equation and ensuring both sides calculate to be equal.

Creating, Solving, or Interpreting Linear Inequalities in One Variable

Linear inequalities and linear equations are both comparisons of two algebraic expressions. However, unlike equations in which the expressions are equal to each other, linear inequalities compare expressions that are unequal. Linear equations typically have one value for the variable that makes the statement true. Linear inequalities generally have an infinite number of values that make the statement true. Exceptions to these last two statements are covered in Section 6.

Writing Linear Inequalities

Linear inequalities are a concise mathematical way to express the relationship between unequal values. More specifically, they describe in what way the values are unequal. A value could be greater than (>); less than (<); greater than or equal to (≥); or less than or equal to (≤) another value. The statement "five times a number added to forty is more than sixty-five" can be expressed as $5x + 40 > 65$. Common words and phrases that express inequalities are:

Symbol	Phrase
<	is under, is below, smaller than, beneath
>	is above, is over, bigger than, exceeds
≤	no more than, at most, maximum
≥	no less than, at least, minimum

Solving Linear Inequalities

When solving a linear inequality, the solution is the set of all numbers that makes the statement true. The inequality $x + 2 \geq 6$ has a solution set of 4 and every number greater than 4 (4.0001, 5, 12, 107, etc.). Adding 2 to 4 or any number greater than 4 would result in a value that is greater than or equal to 6. Therefore, $x \geq 4$ would be the solution set.

Solution sets for linear inequalities often will be displayed using a number line. If a value is included in the set (≥ or ≤), there is a shaded dot placed on that value and an arrow extending in the direction of the solutions. For a variable > or ≥ a number, the arrow would point right on the number line (the direction where the numbers increase); and if a variable is < or ≤ a number, the arrow would point left (where the numbers decrease). If the value is not included in the set (> or <), an open circle on that value would be used with an arrow in the appropriate direction.

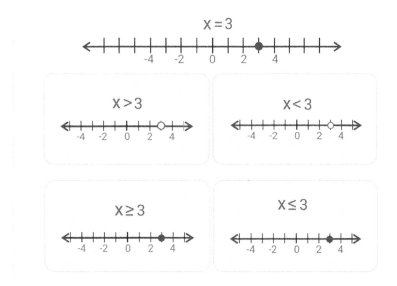

Students may be asked to write a linear inequality given a graph of its solution set. To do so, they should identify whether the value is included (shaded dot or open circle) and the direction in which the arrow is pointing.

In order to algebraically solve a linear inequality, the same steps should be followed as in solving a linear equation (see section on *Solving Linear Equations*). The inequality symbol stays the same for all operations EXCEPT when dividing by a negative number. If dividing by a negative number while solving an inequality, the relationship reverses (the sign flips). Dividing by a positive does not change the relationship, so the sign stays the same. In other words, > switches to < and vice versa. An example is shown below.

Solve $-2(x + 4) \leq 22$

Distribute: $-2x - 8 \leq 22$

Add 8 to both sides: $-2x \leq 30$

Divide both sides by -2: $x \geq 15$

Building a Linear Function that Models a Linear Relationship Between Two Quantities

Linear relationships between two quantities can be expressed in two ways: function notation or as a linear equation with two variables. The relationship is referred to as linear because its graph is represented by a line. For a relationship to be linear, both variables must be raised to the first power only.

Function/Linear Equation Notation

A relation is a set of input and output values that can be written as ordered pairs. A function is a relation in which each input is paired with exactly one output. The domain of a function consists of all inputs, and the range consists of all outputs. Graphing the ordered pairs of a linear function produces a straight line. An example of a function would be $f(x) = 4x + 4$, read "f of x is equal to four times x plus four." In this example, the input would be x and the output would be f(x). Ordered pairs would be represented as (x, f(x)). To find the output for an input value of 3, 3 would be substituted for x into the function as follows: $f(3) = 4(3) + 4$, resulting in $f(3) = 16$. Therefore, the ordered pair $(3, f(3)) = (3, 16)$. Note f(x) is a function of x denoted by f. Functions of x could be named g(x), read "g of x"; p(x), read "p of x"; etc.

A linear function could also be written in the form of an equation with two variables. Typically, the variable x represents the inputs and the variable y represents the outputs. The variable x is considered the independent variable and y the dependent variable. The above function would be written as $y = 4x + 4$. Ordered pairs are written in the form (x, y).

Writing Linear Equations in Two Variables

When writing linear equations in two variables, the process depends on the information given. Questions will typically provide the slope of the line and its y-intercept, an ordered pair and the slope, or two ordered pairs.

Given the Slope and Y-Intercept

Linear equations are commonly written in slope-intercept form, $y = mx + b$, where m represents the slope of the line and b represents the y-intercept. The slope is the rate of change between the variables, usually expressed as a whole number or fraction. The y-intercept is the value of y when x = 0 (the point where the line intercepts the y-axis on a graph). Given the slope and y-intercept of a line, the values are

68

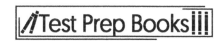

substituted for *m* and *b* into the equation. A line with a slope of ½ and *y*-intercept of -2 would have an equation $y = \frac{1}{2}x - 2$.

Given an Ordered Pair and the Slope
The point-slope form of a line, $y - y_1 = m(x - x_1)$, is used to write an equation when given an ordered pair (point on the equation's graph) for the function and its rate of change (slope of the line). The values for the slope, *m*, and the point (x_1, y_1) are substituted into the point-slope form to obtain the equation of the line. A line with a slope of 3 and an ordered pair (4, -2) would have an equation $y - (-2) = 3(x - 4)$. If a question specifies that the equation be written in slope-intercept form, the equation should be manipulated to isolate *y*: ~~Turning point-slope form into slope-intercept-~~

Solve: $y - (-2) = 3(x - 4)$

Distribute: $y + 2 = 3x - 12$

Subtract 2 from both sides: $y = 3x - 14$

Given Two Ordered Pairs
Given two ordered pairs for a function, (x_1, y_1) and (x_2, y_2), it is possible to determine the rate of change between the variables (slope of the line). To calculate the slope of the line, m, the values for the ordered pairs should be substituted into the formula:

$$m = \frac{y_2 - y_1}{x_2 - x_1}$$

The expression is substituted to obtain a whole number or fraction for the slope. Once the slope is calculated, the slope and either of the ordered pairs should be substituted into the point-slope form to obtain the equation of the line.

Creating, Solving, and Interpreting Systems of Linear Inequalities in Two Variables

Expressing Linear Inequalities in Two Variables
A linear inequality in two variables is a statement expressing an unequal relationship between those two variables. Typically written in slope-intercept form, the variable *y* can be greater than; less than; greater than or equal to; or less than or equal to a linear expression including the variable *x*. Examples include $y > 3x$ and $y \leq \frac{1}{2}x - 3$. Questions may instruct students to model real world scenarios such as:

> You work part-time cutting lawns for $15 each and cleaning houses for $25 each. Your goal is to make more than $90 this week. Write an inequality to represent the possible pairs of lawns and houses needed to reach your goal.

This scenario can be expressed as $15x + 25y > 90$ where *x* is the number of lawns cut and *y* is the number of houses cleaned.

Graphing Solution Sets for Linear Inequalities in Two Variables
A graph of the solution set for a linear inequality shows the ordered pairs that make the statement true. The graph consists of a boundary line dividing the coordinate plane and shading on one side of the boundary. The boundary line should be graphed just as a linear equation would be graphed (see section on *Understanding Connections Between Algebraic and Graphical Representations*). If the inequality symbol is > or <, a dashed line can be used to indicate that the line is not part of the solution set. If the

inequality symbol is ≥ or ≤, a solid line can be used to indicate that the boundary line is included in the solution set. An ordered pair (x, y) on either side of the line should be chosen to test in the inequality statement. If substituting the values for x and y results in a true statement $(15(3) + 25(2) > 90)$, that ordered pair and all others on that side of the boundary line are part of the solution set. To indicate this, that region of the graph should be shaded. If substituting the ordered pair results in a false statement, the ordered pair and all others on that side are not part of the solution set.

Therefore, the other region of the graph contains the solutions and should be shaded.

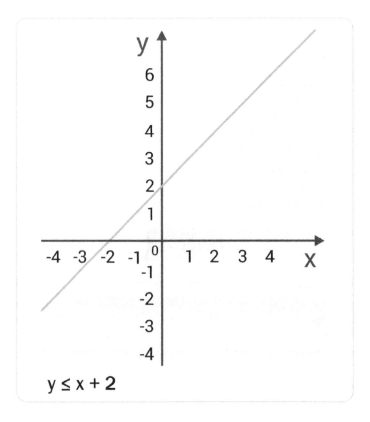

$$y \leq x + 2$$

A question may simply ask whether a given ordered pair is a solution to a given inequality. To determine this, the values should be substituted for the ordered pair into the inequality. If the result is a true statement, the ordered pair is a solution; if the result is a false statement, the ordered pair is not a solution.

Expressing Systems of Linear Inequalities in Two Variables

A system of linear inequalities consists of two linear inequalities making comparisons between two variables. Students may be given a scenario and asked to express it as a system of inequalities:

> A consumer study calls for at least 60 adult participants. It cannot use more than 25 men. Express these constraints as a system of inequalities.

This can be modeled by the system: $x + y \geq 60; x \leq 25$, where x represents the number of men and y represents the number of women. A solution to the system is an ordered pair that makes both inequalities true when substituting the values for x and y.

Graphing Solution Sets for Systems of Linear Inequalities in Two Variables

The solution set for a system of inequalities is the region of a graph consisting of ordered pairs that make both inequalities true. To graph the solution set, each linear inequality should first be graphed with appropriate shading. The region of the graph should be identified where the shading for the two inequalities overlaps. This region contains the solution set for the system.

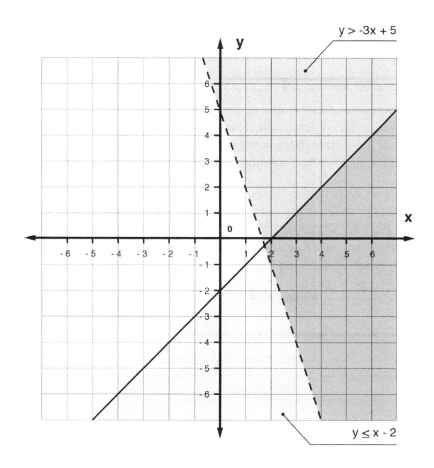

An ordered pair from the region of solutions can be selected to test in the system of inequalities.

Just as with manipulating linear inequalities in one variable, if dividing by a negative number in working with a linear inequality in two variables, the relationship reverses and the inequality sign should be flipped.

Creating, Solving, and Interpreting Systems of Two Linear Equations in Two Variables

Expressing Systems of Two Linear Equations in Two Variables

A system of two linear equations in two variables is a set of equations that use the same variables, usually x and y. Here's a sample problem:

> An Internet provider charges an installation fee and a monthly charge. It advertises that two months of its offering costs $100 and six months costs $200. Find the monthly charge and the installation fee.

The two unknown quantities (variables) are the monthly charge and the installation fee. There are two different statements given relating the variables: two months added to the installation fee is $100; and six months added to the installation fee is $200. Using the variable x as the monthly charge and y as the installation fee, the statements can be written as the following: $2x + y = 100$; $6x + y = 200$. These two equations taken together form a system modeling the given scenario.

Solutions of a System of Two Linear Equations in Two Variables

A solution for a system of equations is an ordered pair that makes both equations true. One method for solving a system of equations is to graph both lines on a coordinate plane (see section on *Understanding Connections Between Algebraic and Graphical Representations*). If the lines intersect, the point of intersection is the solution to the system. Every point on a line represents an ordered pair that makes its equation true. The ordered pair represented by this point of intersection lies on both lines and therefore makes both equations true. This ordered pair should be checked by substituting its values into both of the original equations of the system. Note that given a system of equations and an ordered pair, the ordered pair can be determined to be a solution or not by checking it in both equations.

If, when graphed, the lines representing the equations of a system do not intersect, then the two lines are parallel to each other or they are the same exact line. Parallel lines extend in the same direction without ever meeting. A system consisting of parallel lines has no solution. If the equations for a system represent the same exact line, then every point on the line is a solution to the system. In this case, there would be an infinite number of solutions. A system consisting of intersecting lines is referred to as independent; a system consisting of parallel lines is referred to as inconsistent; and a system consisting of coinciding lines is referred to as dependent.

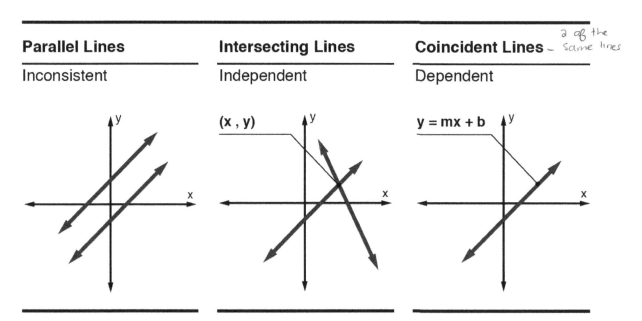

Parallel Lines	**Intersecting Lines**	**Coincident Lines** — 2 of the same lines
Inconsistent	Independent	Dependent

Algebraically Solving Linear Equations (or Inequalities) in One Variable

Linear equations in one variable and linear inequalities in one variable can be solved following similar processes. Although they typically have one solution, a linear equation can have no solution or can have a solution set of all real numbers. Solution sets for linear inequalities typically consist of an infinite number of values either greater or less than a given value (where the given value may or may not be

included in the set). However, a linear inequality can have no solution or can have a solution set consisting of all real numbers.

Linear Equations in One Variable – Special Cases

Solving a linear equation produces a value for the variable that makes the algebraic statement true. If there is no value for the variable that would make the statement true, there is no solution to the equation. Here's a sample equation: $x + 3 = x - 1$. There is no value for x in which adding 3 to the value would produce the same result as subtracting 1 from that value. Conversely, if any value for the variable would make a true statement, the equation has an infinite number of solutions. Here's another sample equation: $3x + 6 = 3(x + 2)$. Any real number substituted for x would result in a true statement (both sides of the equation are equal).

By manipulating equations similar to the two above, the variable of the equation will cancel out completely. If the constants that are left express a true statement (ex., $6 = 6$), then all real numbers are solutions to the equation. If the constants left express a false statement (ex., $3 = -1$), then there is no solution to the equation.

A question on this material may present a linear equation with an unknown value for either a constant or a coefficient of the variable and ask to determine the value that produces an equation with no solution or infinite solutions. For example:

$3x + 7 = 3x + 10 + n$; Find the value of n that would create an equation with an infinite number of solutions for the variable x.

To solve this problem, the equation should be manipulated so the variable x will cancel. To do this, $3x$ should be subtracted from both sides, which would leave $7 = 10 + n$. By subtracting 10 on both sides, it is determined that $n = -3$. Therefore, a value of -3 for n would result in an equation with a solution set of all real numbers.

If the same problem asked for the equation to have no solution, the value of n would be all real numbers except -3.

Linear Inequalities in One Variable – Special Cases

A linear inequality can have a solution set consisting of all real numbers or can contain no solution. When solved algebraically, a linear inequality in which the variable cancels out and results in a true statement (ex., $7 \geq 2$) has a solution set of all real numbers. A linear inequality in which the variable cancels out and results in a false statement (ex., $7 \leq 2$) has no solution.

Compound Inequalities

A compound inequality is a pair of inequalities joined by *and* or *or*. Given a compound inequality, to determine its solution set, both inequalities should be solved for the given variable. The solution set for a compound inequality containing *and* consists of all the values for the variable that make both inequalities true. If solving the compound inequality results in $x > -9$ and $x \leq 6$, the solution set would consist of all values between -2 and 3, including 3. This may also be written as follows: $-9 < x \leq 6$. Due to the graphs of their solution sets (shown below), compound inequalities such as these are referred to as conjunctions.

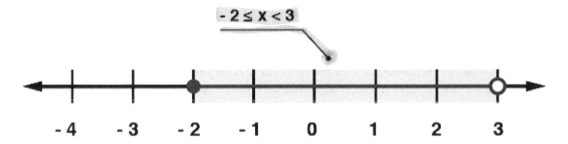

If there are no values that would make both inequalities of a compound inequality containing *and* true, then there is no solution. An example would be $x > 2$ and $x \leq 0$.

The solution set for a compound inequality containing *or* consists of all the values for the variable that make at least one of the inequalities true. The solution set for the compound inequality $x < 3$ or $x \geq 6$ consists of all values less than 3, 6, and all values greater than 6. Due to the graphs of their solution sets (shown below), compound inequalities such as these are referred to as disjunctions.

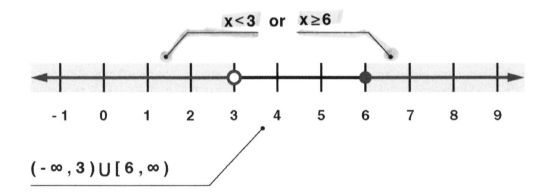

If the two inequalities for a compound inequality containing *or* "overlap," then the solution set contains all real numbers. An example would be $x > 2$ or $x < 7$. Any number would make at least one of these true.

Algebraically Solving Systems of Two Linear Equations in Two Variables

A system of two linear equations in two variables is a set of equations that use the same variables (typically x and y). A solution to the system is an ordered pair that makes both equations true. One method for solving a system is by graphing as explained in Section 5.2. This method, however, is not always practical. Students may not have graph paper; or the solution may not consist of integers, making it difficult to identify the exact point of intersection on a graph. There are two methods for solving systems of equations algebraically: substitution and elimination. The method used will depend on the characteristics of the equations in the system.

Solving Systems of Equations with the Substitution Method

If one of the equations in a system has an isolated variable ($x=$ or $y=$) or a variable that can be easily isolated, the substitution method can be used. Here's a sample system: $x + 3y = 7; 2x - 4y = 24$. The

first equation can easily be solved for x. By subtracting 3y on both sides, the resulting equation is $x = 7 - 3y$. When one equation is solved for a variable, the expression that it is equal can be substituted into the other equation. For this example, $(7 - 3y)$ would be substituted for x into the second equation as follows: $2(7 - 3y) + 4y = 24$. Solving this equation results in $y = -5$. Once the value for one variable is known, this value should be substituted into either of the original equations to determine the value of the other variable. For the example, -5 would be substituted for y in either of the original equations. Substituting into the first equation results in $x + 3(-5) = 7$, and solving this equation yields $x = 22$. The solution to a system is an ordered pair, so the solution to the example is written as (22, 7). The solution can be checked by substituting it into both equations of the system to ensure it results in two true statements.

Solving Systems of Equations with the Elimination Method

The elimination method for solving a system of equations involves canceling out (or eliminating) one of the variables. This method is typically used when both equations of a system are written in standard form $(Ax + By = C)$. An example is $2x + 3y = 12; 5x - y = 13$. To perform the elimination method, the equations in the system should be arranged vertically to be added together and then one or both of the equations should be multiplied so that one variable will be eliminated when the two are added. Opposites will cancel each other when added together. For example, 8x and -8x will cancel each other when added. For the example above, writing the system vertically helps identify that the bottom equation should be multiplied by 3 to eliminate the variable y.

$$2x + 3y = 12 \quad \rightarrow \quad 2x + 3y = 12$$

$$3(5x - y = 13) \quad \rightarrow \quad 15x - 3y = 39$$

Adding the two equations together vertically results in $17x = 51$. Solving yields $x = 3$. Once the value for one variable is known, it can be substituted into either of the original equations to determine the value of the other variable. Once this is obtained, the solution can be written as an ordered pair (x, y) and checked in both equations of the system. In this example, the solution is (3, 2).

Systems of Equations with No Solution or an Infinite Number of Solutions

A system of equations can have one solution, no solution, or an infinite number of solutions. If, while solving a system algebraically, both variables cancel out, then the system has either no solution or has an infinite number of solutions. If the remaining constants result in a true statement (ex., $7 = 7$), then there is an infinite number of solutions. This would indicate coinciding lines. If the remaining constants result in a false statement, then there is no solution to the system. This would indicate parallel lines.

Interpreting Variables and Constants in Expressions for Linear Functions in the Context Presented

Linear functions, also written as linear equations in two variables, can be written to model real-world scenarios. Questions on this material will provide information about a scenario and then request a linear equation to represent the scenario. The algebraic process for writing the equation will depend on the given information. The key to writing linear models is to decipher the information given to determine what it represents in the context of a linear equation (variables, slope, ordered pairs, etc.).

Identifying Variables for Linear Models

The first step to writing a linear model is to identify what the variables represent. A variable represents an unknown quantity, and in the case of a linear equation, a specific relationship exists between the two variables (usually *x* and *y*). Within a given scenario, the variables are the two quantities that are changing. The variable *x* is considered the independent variable and represents the inputs of a function. The variable *y* is considered the dependent variable and represents the outputs of a function. For example, if a scenario describes distance traveled and time traveled, distance would be represented by *y* and time represented by *x*. The distance traveled depends on the time spent traveling (time is independent). If a scenario describes the cost of a cab ride and the distance traveled, the cost would be represented by *y* and the distance represented by *x*. The cost of a cab ride depends on the distance traveled.

Identifying the Slope and Y-Intercept for Linear Models

The slope of the graph of a line represents the rate of change between the variables of an equation. In the context of a real-world scenario, the slope will tell the way in which the unknown quantities (variables) change with respect to each other. A scenario involving distance and time might state that someone is traveling at a rate of 45 miles per hour. The slope of the linear model would be 45. A scenario involving the cost of a cab ride and distance traveled might state that the person is charged $3 for each mile. The slope of the linear model would be 3.

The *y*-intercept of a linear function is the value of *y* when $x = 0$ (the point where the line intercepts the *y*-axis on the graph of the equation). It is sometimes helpful to think of this as a "starting point" for a linear function. Suppose for the scenario about the cab ride that the person is told that the cab company charges a flat fee of $5 plus $3 for each mile. Before traveling any distance ($x = 0$), the cost is $5. The *y*-intercept for the linear model would be 5.

Identifying Ordered Pairs for Linear Models

A linear equation with two variables can be written given a point (ordered pair) and the slope or given two points on a line. An ordered pair gives a set of corresponding values for the two variables (*x* and *y*). As an example, for a scenario involving distance and time, it is given that the person traveled 112.5 miles in 2 ½ hours. Knowing that *x* represents time and *y* represents distance, this information can be written as the ordered pair (2.5, 112.5).

Understanding Connections Between Algebraic and Graphical Representations

The solution set to a linear equation in two variables can be represented visually by a line graphed on the coordinate plane. Every point on this line represents an ordered pair (*x*, *y*), which makes the equation true. The process for graphing a line depends on the form in which its equation is written: slope-intercept form or standard form.

Graphing a Line in Slope-Intercept Form

When an equation is written in slope-intercept form, $y = mx + b$, *m* represents the slope of the line and *b* represents the *y*-intercept. The *y*-intercept is the value of *y* when $x = 0$ and the point at which the graph of the line crosses the *y*-axis. The slope is the rate of change between the variables, expressed as a fraction. The fraction expresses the change in *y* compared to the change in *x*. If the slope is an integer, it should be written as a fraction with a denominator of 1. For example, 5 would be written as 5/1.

To graph a line given an equation in slope-intercept form, the *y*-intercept should first be plotted. For example, to graph $y = -\frac{2}{3}x + 7$, the *y*-intercept of 7 would be plotted on the *y*-axis (vertical axis) at the point (0, 7). Next, the slope would be used to determine a second point for the line. Note that all that is

necessary to graph a line is two points on that line. The slope will indicate how to get from one point on the line to another.

The slope expresses vertical change (*y*) compared to horizontal change (*x*) and therefore is sometimes referred to as:

$$\frac{rise}{run}$$

The numerator indicates the change in the *y* value (move up for positive integers and move down for negative integers), and the denominator indicates the change in the *x* value. For the previous example, using the slope of $-\frac{2}{3}$, from the first point at the *y*-intercept, the second point should be found by counting down 2 and to the right 3. This point would be located at (3, 5).

Graphing a Line in Standard Form

When an equation is written in standard form, $Ax + By = C$, it is easy to identify the *x*- and *y*-intercepts for the graph of the line. Just as the *y*-intercept is the point at which the line intercepts the *y*-axis, the *x*-intercept is the point at which the line intercepts the *x*-axis. At the *y*-intercept, $x = 0$; and at the *x*-intercept, $y = 0$. Given an equation in standard form, $x = 0$ should be used to find the *y*-intercept. Likewise, $y = 0$ should be used to find the *x*-intercept. For example, to graph $3x + 2y = 6$, 0 for *y* results in $3x + 2(0) = 6$. Solving for *y* yields $x = 2$; therefore, an ordered pair for the line is (2, 0). Substituting 0 for *x* results in $3(0) + 2y = 6$. Solving for *y* yields $y = 3$; therefore, an ordered pair for the line is (0, 3). The two ordered pairs (the *x*- and *y*-intercepts) can be plotted, and a straight line through them can be constructed.

or you can convert it to slope intercept form and graph

T - chart			Intercepts
x	**y**		x - intercept : (2,0)
0	3		y - intercept : (0,3)
2	0		

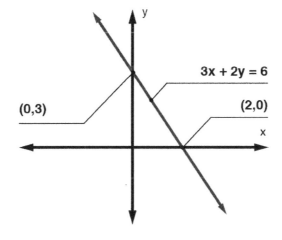

Writing the Equation of a Line Given its Graph

Given the graph of a line, its equation can be written in two ways. If the y-intercept is easily identified (is an integer), it and another point can be used to determine the slope. When determining $\frac{change\ in\ y}{change\ in\ x}$ from one point to another on the graph, the distance for $\frac{rise}{run}$ is being figured. The equation should be written in slope-intercept form, $y = mx + b$, with m representing the slope and b representing the y-intercept.

The equation of a line can also be written by identifying two points on the graph of the line. To do so, the slope is calculated and then the values are substituted for the slope and either of the ordered pairs into the point-slope form of an equation.

Vertical, Horizontal, Parallel, and Perpendicular Lines

For a vertical line, the value of x remains constant (for all ordered pairs (x, y) on the line, the value of x is the same); therefore, the equations for all vertical lines are written in the form $x = number$. For example, a vertical line that crosses the x-axis at -2 would have an equation of $x = -2$. For a horizontal line, the value of y remains constant; therefore, the equations for all horizontal lines are written in the form $y = number$.

Parallel lines extend in the same exact direction without ever meeting. Their equations have the same slopes and different y-intercepts. For example, given a line with an equation of $y = -3x + 2$, a parallel line would have a slope of -3 and a y-intercept of any value other than 2. Perpendicular lines intersect to form a right angle. Their equations have slopes that are opposite reciprocal (the sign is changed and the fraction is flipped; for example, $-\frac{2}{3}$ and $\frac{3}{2}$) and y-intercepts that may or may not be the same. For example, given a line with an equation of $y = \frac{1}{2}x + 7$, a perpendicular line would have a slope of $-\frac{2}{1}$ and any value for its y-intercept.

Problem-Solving and Data Analysis

Using Ratios, Rates, Proportions, and Scale Drawings to Solve Single- and Multistep Problems

Ratios, rates, proportions, and scale drawings are used when comparing two quantities. Questions on this material will include expressing relationships in simplest terms and solving for missing quantities.

Ratios

A ratio is a comparison of two quantities that represent separate groups. For example, if a recipe calls for 2 eggs for every 3 cups of milk, it can be expressed as a ratio. Ratios can be written three ways: (1) with the word "to"; (2) using a colon; or (3) as a fraction. For the previous example, the ratio of eggs to cups of milk can be written as: 2 to 3, 2:3, or $\frac{2}{3}$. When writing ratios, the order is important. The ratio of eggs to cups of milk is not the same as the ratio of cups of milk to eggs, 3:2.

In simplest form, both quantities of a ratio should be written as integers. These should also be reduced just as a fraction would be. For example, 5:10 would reduce to 1:2. Given a ratio where one or both quantities are expressed as a decimal or fraction, both should be multiplied by the same number to produce integers. To write the ratio $\frac{1}{3}$ to 2 in simplest form, both quantities should be multiplied by 3. The resulting ratio is 1 to 6.

When a problem involving ratios gives a comparison between two groups, then: (1) a total should be provided and a part should be requested; or (2) a part should be provided and a total should be requested. Consider the following:

The ratio of boys to girls in the 11th grade is 5:4. If there is a total of 270 11th grade students, how many are girls?

To solve this, the total number of "ratio pieces" first needs to be determined. The total number of 11th grade students is divided into 9 pieces. The ratio of boys to total students is 5:9; and the ratio of girls to total students is 4:9. Knowing the total number of students, the number of girls can be determined by setting up a proportion:

$$\frac{4}{9} = \frac{x}{270}$$

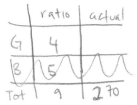

Solving the proportion, it shows that there are 120 11th grade girls.

Rates
A rate is a ratio comparing two quantities expressed in different units. A unit rate is one in which the second is one unit. Rates often include the word *per*. Examples include miles per hour, beats per minute, and price per pound. The word *per* can be represented with a / symbol or abbreviated with the letter "p" and the units abbreviated. For example, miles per hour would be written mi/h. Given a rate that is not in simplest form (second quantity is not one unit), both quantities should be divided by the value of the second quantity. Suppose a patient had 99 heartbeats in 1½ minutes. To determine the heart rate, 1½ should divide both quantities. The result is 66 bpm.

Scale Drawings
Scale drawings are used in designs to model the actual measurements of a real-world object. For example, the blueprint of a house might indicate that it is drawn at a scale of 3 inches to 8 feet. Given one value and asked to determine the width of the house, a proportion should be set up to solve the problem. Given the scale of 3in:8ft and a blueprint width of 1 ft (12 in.), to find the actual width of the building, the proportion $\frac{3}{8} = \frac{12}{x}$ should be used. This results in an actual width of 32 ft.

Proportions
A proportion is a statement consisting of two equal ratios. Proportions will typically give three of four quantities and require solving for the missing value. The key to solving proportions is to set them up properly. Here's a sample problem:

If 7 gallons of gas costs $14.70, how many gallons can you get for $20?

The information should be written as equal ratios with a variable representing the missing quantity:

$$\left(\frac{gallons}{cost} = \frac{gallons}{cost}\right) : \frac{7}{14.70} = \frac{x}{20}$$

To solve, cross multiply (multiply the numerator of the first ratio by the denominator of the second and vice versa) is used and the products are set equal to each other. Cross-multiplying results in: $(7)(20) = (14.7)(x)$. Solving the equation for x, it can be determined that 9.5 gallons of gas can be purchased for $20.

Indirect Proportions

The proportions described above are referred to as direct proportions or direct variation. For direct proportions, as one quantity increases, the other quantity also increases. For indirect proportions (also referred to as indirect variations, inverse proportions, or inverse variations), as one quantity increases, the other decreases. Direct proportions can be written:

$$\text{direct prop}: \quad \frac{y_1}{x_1} = \frac{y_2}{x_2}$$

Conversely, indirect proportions are written:

$$\text{indirect prop}: \quad y_1 x_1 = y_2 x_2$$

Here's a sample problem:

> It takes 3 carpenters 10 days to build the frame of a house. How long should it take 5 carpenters to build the same frame?

In this scenario, as one quantity increases (number of carpenters), the other decreases (number of days building); therefore, this is an inverse proportion. To solve, the products of the two variables (in this scenario, the total work performed) are set equal to each other:

$$(y_1 x_1 = y_2 x_2)$$

Using *y* to represent carpenters and *x* to represent days, the resulting equation is: $(3)(10) = (5)(x2)$. Solving for x_2, it is determined that it should take 5 carpenters 6 days to build the frame of the house.

Solving Single- and Multistep Problems Involving Percentages

The word percent means "per hundred." When dealing with percentages, it may be helpful to think of the number as a value in hundredths. For example, 15% can be expressed as "fifteen hundredths" and written as $\frac{15}{100}$ or .15.

Converting from Decimals and Fractions to Percentages

To convert a decimal to a percent, a number is multiplied by 100. To write .25 as a percent, the equation $.25 x 100$ yields 25%. To convert a fraction to a percent, the fraction is converted to a decimal and then multiplied by 100. To convert $\frac{3}{5}$ to a decimal, the numerator (3) is divided by the denominator (5). This results in .6, which is then multiplied by 100 to get 60%.

To convert a percent to a decimal, the number is divided by 100. For example, 150% is equal to 1.5 $\left(\frac{150}{100}\right)$. To convert a percent to a fraction, the percent sign is deleted and the value is written as the numerator with a denominator of 100. For example, 2% = $\frac{2}{100}$. Fractions should be reduced: $\frac{2}{100} = \frac{1}{50}$.

Percent Problems

Material on percentages can include questions such as: What is 15% of 25? What percent of 45 is 3? Five is $\frac{1}{2}$% of what number? To solve these problems, the information should be rewritten as an equation where the following helpful steps are completed: (1) "what" is represented by a variable (*x*); (2) "is" is represented by an = sign; and (3) "of" is represented by multiplication. Any values expressed as a

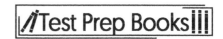

percent should be written as a decimal; and if the question is asking for a percent, the answer should be converted accordingly. Here are three sample problems based on the information above:

What is 15% of 25?
$$x = .15 \times 25$$
$$x = 3.75$$

What percent of 45 is 3?
$$x \times 45 = 3$$
$$x = 0.0\overline{6}$$
$$x = 6.\overline{6}\%$$

Five is $\frac{1}{2}$% of what number?
$$5 = .005 \times x$$
$$x = 1,000$$

Percent Increase/Decrease
Problems dealing with percentages may involve an original value, a change in that value, and a percentage change. A problem will provide two pieces of information and ask to find the third. To do so, this formula is used:

$$\frac{change}{original\ value} \times 100 = percent\ change$$

Here's a sample problem:

Attendance at a baseball stadium has dropped 16% from last year. Last year's average attendance was 40,000. What is this year's average attendance?

Using the formula and information, the change is unknown (x), the original value is 40,000, and the percent change is 16%. The formula can be written as:

$$\frac{x}{40,000} \times 100 = 16$$

When solving for x, it is determined the change was 6,400. The problem asked for this year's average attendance, so to calculate, the change (6,400) is subtracted from last year's attendance (40,000) to determine this year's average attendance is 33,600.

Percent More Than/Less Than
Percentage problems may give a value and what percent that given value is more than or less than an original unknown value. Here's a sample problem:

A store advertises that all its merchandise has been reduced by 25%. The new price of a pair of shoes is $60. What was the original price?

This problem can be solved by writing a proportion. Two ratios should be written comparing the cost and the percent of the original cost. The new cost is 75% of the original cost (100% - 25%); and the original cost is 100% of the original cost. The unknown original cost can be represented by x. The proportion would be set up as:

$$\frac{60}{75} = \frac{x}{100}$$

Solving the proportion, it is determined the original cost was $80.

81

Solving Single- and Multistep Problems Involving Measurement Quantities, Units, and Unit Conversion

Unit Rates

A rate is a ratio in which two terms are in different units. When rates are expressed as a quantity of one, they are considered unit rates. To determine a unit rate, the first quantity is divided by the second. Knowing a unit rate makes calculations easier than simply having a rate. Suppose someone bought a 3lb bag of onions for $1.77. To calculate the price of 5lbs of onions, a proportion could be set up as follows:

$$\frac{3}{1.77} = \frac{5}{x}$$

However, knowing the unit rate, multiplying the value of pounds of onions by the unit price is another way to find the solution: (The unit price would be calculated $1.77/3lb = $0.59/lb.)

$$5lbs \times \frac{\$.59}{lb} = \$2.95$$

(The "lbs" units cancel out.)

Unit Conversion

Unit conversions apply to many real-world scenarios, including cooking, measurement, construction, and currency. Problems on this material can be solved similarly to those involving unit rates. Given the conversion rate, it can be written as a fraction (ratio) and multiplied by a quantity in one unit to convert it to the corresponding unit. For example, someone might want to know how many minutes are in 3½ hours. The conversion rate of 60 minutes to 1 hour can be written as $\frac{60 \ min}{1 \ h}$. Multiplying the quantity by the conversion rate results in:

$$3\frac{1}{2}h \times \frac{60 \ min}{1 \ h} = 210 \ min$$

The "h" unit is canceled. To convert a quantity in minutes to hours, the fraction for the conversion rate would be flipped (to cancel the "min" unit). To convert 195 minutes to hours, the equation $195 \ min \times \frac{1h}{60min}$ would be used. The result is $\frac{195h}{60}$, which reduces to $3\frac{1}{4}$ hours.

Converting units may require more than one multiplication. The key is to set up the conversion rates so that units cancel out each other and the desired unit is left. Suppose someone wants to convert 3.25 yards to inches, given that 1yd = 3ft and 12in = 1ft. To calculate, the equation:

$$3.25yd \times \frac{3ft}{1yd} \times \frac{12in}{1ft}$$
would be used

The "yd" and "ft" units will cancel, resulting in 117 inches.

Given a Scatterplot, Using Linear, Quadratic, or Exponential Models to Describe How Variables are Related

Scatterplots can be used to determine whether a correlation exists between two variables. The horizontal (*x*) axis represents the independent variable and the vertical (*y*) axis represents the dependent variable. If when graphed, the points model a linear, quadratic, or exponential relationship, then a correlation is said to exist. If so, a line of best-fit or curve of best-fit can be drawn through the points, with the points relatively close on either side. Writing the equation for the line or curve allows for predicting values for the variables. Suppose a scatterplot displays the value of an investment as a function of years after investing. By writing an equation for the line or curve and substituting a value for one variable into the equation, the corresponding value for the other variable can be calculated.

Linear Models

If the points of a scatterplot model a linear relationship, a line of best-fit is drawn through the points. If the line of best-fit has a positive slope (*y*-values increase as *x*-values increase), then the variables have a positive correlation. If the line of best-fit has a negative slope (*y*-values decrease as *x*-values increase), then a negative correlation exists. A positive or negative correlation can also be categorized as strong or weak, depending on how closely the points are grouped around the line of best-fit.

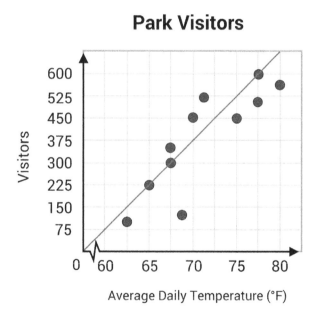

Park Visitors

Given a line of best-fit, its equation can be written by identifying: the slope and *y*-intercept; a point and the slope; or two points on the line.

Quadratic Models

A quadratic function can be written in the form $y = ax^2 + bx + c$. The u-shaped graph of a quadratic function is called a parabola. The graph can either open up or open down (upside down u). The graph is symmetric about a vertical line, called the axis of symmetry. Corresponding points on the parabola are directly across from each other (same *y*-value) and are the same distance from the axis of symmetry (on either side). The axis of symmetry intersects the parabola at its vertex. The *y*-value of the vertex represents the minimum or maximum value of the function. If the graph opens up, the value of *a* in its equation is positive, and the vertex represents the minimum of the function.

If the graph opens down, the value of *a* in its equation is negative, and the vertex represents the maximum of the function.

find vertex:
- $\frac{-b}{2a} = x$
- (h, k)

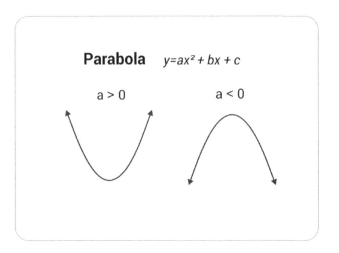

Given a curve of best-fit that models a quadratic relationship, the equation of the parabola can be written by identifying the vertex of the parabola and another point on the graph. The values for the vertex (h, k) and the point (x, y) should be substituted into the vertex form of a quadratic function:

$$y = a(x - h)^2 - k$$

to determine the value of *a*. To write the equation of a quadratic function with a vertex of (4, 7) and containing the point (8, 3), the values for *h*, *k*, *x*, and *y* should be substituted into the vertex form of a quadratic function, resulting in:

$$3 = a(8 - 4)^2 + 7$$

Solving for *a*, yields $a = -\frac{1}{4}$. Therefore, the equation of the function can be written as:

$$y = -\frac{1}{4}(x - 4)^2 + 7$$

The vertex form can be manipulated in order to write the quadratic function in standard form.

Exponential Models

An exponential curve can be used as a curve of best-fit for a scatterplot. The general form for an exponential function is $y = ab^x$ where b must be a positive number and cannot equal 1. When the value of b is greater than 1, the function models exponential growth (as x increases, y increases). When the value of b is less than 1, the function models exponential decay (as x increases, y decreases). If a is positive, the graph consists of points above the x-axis; and if a is negative, the graph consists of points below the x-axis. An asymptote is a line that a graph approaches.

Exponential Curve

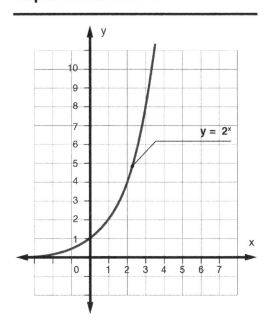

Given a curve of best-fit modeling an exponential function, its equation can be written by identifying two points on the curve. To write the equation of an exponential function containing the ordered pairs (2, 2) and (3, 4), the ordered pair (2, 2) should be substituted in the general form and solved for a:

$$2 = a \times b^2 \rightarrow a = \frac{2}{b^2}$$

The ordered pair (3, 4) and $\frac{2}{b^2}$ should be substituted in the general form and solved for b:

$$4 = \frac{2}{b^2} \times b^3 \rightarrow b = 2$$

Then, 2 should be substituted for b in the equation for a and then solved for a:

$$a = \frac{2}{2^2} \rightarrow a = \frac{1}{2}$$

Knowing the values of a and b, the equation can be written as:

$$y = \frac{1}{2} \times 2^x$$

85

Using the Relationship Between Two Variables to Investigate Key Features of a Graph

Material on graphing relationships between two variables may include linear, quadratic, and exponential functions. Graphing linear functions is covered in Section 9.

Graphing Quadratic Functions

The standard form of a quadratic function is:

$$y = ax^2 + bx + c$$

The graph of a quadratic function is a u-shaped (or upside down u) curve, called a parabola, which is symmetric about a vertical line (axis of symmetry). To graph a parabola, its vertex (high or low point for the curve) and at least two points on each side of the axis of symmetry need to be determined.

Given a quadratic function in standard form, the axis of symmetry for its graph is the line $x = -\frac{b}{2a}$. The vertex for the parabola has an x-coordinate of $-\frac{b}{2a}$. To find the y-coordinate for the vertex, the calculated x-coordinate needs to be substituted. To complete the graph, two different x-values need to be selected and substituted into the quadratic function to obtain the corresponding y-values. This will give two points on the parabola. These two points and the axis of symmetry are used to determine the two points corresponding to these. The corresponding points are the same distance from the axis of symmetry (on the other side) and contain the same y-coordinate. Plotting the vertex and four other points on the parabola allows for constructing the curve.

← always include x = ___

Quadratic Function

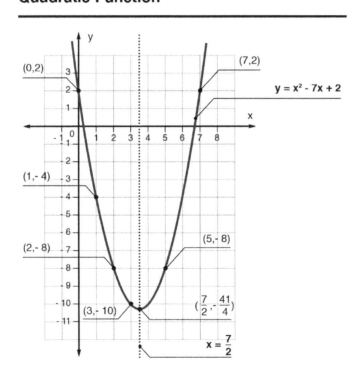

Graphing Exponential Functions

Exponential functions have a general form of $y = a \times b^x$. The graph of an exponential function is a curve that slopes upward or downward from left to right. The graph approaches a line, called an asymptote, as x or y increases or decreases. To graph the curve for an exponential function, x-values are selected and then substituted into the function to obtain the corresponding y-values. A general rule of thumb is to select three negative values, zero, and three positive values. Plotting the seven points on the graph for an exponential function should allow for constructing a smooth curve through them.

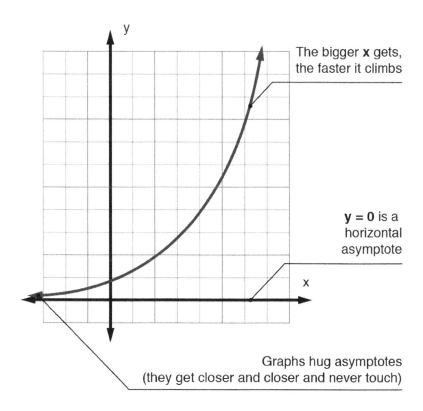

The bigger **x** gets, the faster it climbs

y = 0 is a horizontal asymptote

Graphs hug asymptotes
(they get closer and closer and never touch)

Comparing Linear Growth with Exponential Growth

Both linear and exponential equations can model a relationship of growth or decay between two variables. If the dependent variable (y) increases as the independent variable (x) increases, the relationship is referred to as growth. If y decreases as x increases, the relationship is referred to as decay.

Linear Growth and Decay

A linear function can be written in the form $y = mx + b$, where x represents the inputs, y represents the outputs, b represents the y-intercept for the graph, and m represents the slope of the line. The y-intercept is the value of y when $x = 0$ and can be thought of as the "starting point." The slope is the rate of change between the variables x and y. A positive slope represents growth; and a negative slope represents decay. Given a table of values for inputs (x) and outputs (y), a linear function would model the relationship if: x and y change at a constant rate per unit interval—for every two inputs a given

distance apart, the distance between their corresponding outputs is constant. Here are some sample ordered pairs:

x	0	1	2	3
y	-7	-4	-1	2

For every 1 unit increase in x, y increases by 3 units. Therefore, the change is constant and thus represents linear growth.

Given a scenario involving growth or decay, determining if there is a constant rate of change between inputs (x) and outputs (y) will identify if a linear model is appropriate. A scenario involving distance and time might state that someone is traveling at a rate of 45 miles per hour. For every hour traveled (input), the distance traveled (output) increases by 45 miles. This is a constant rate of change.

The process for writing the equation to represent a linear model is covered in the section *Writing Linear Equations in Two Variables*.

Exponential Growth and Decay

An exponential function can be written in the form $y = a \times b^x$, where x represents the inputs, y represents the outputs, a represents the y-intercept for the graph, and b represents the growth rate. The y-intercept is the value of y when $x = 0$ and can be thought of as the "starting point." If b is greater than 1, the function describes exponential growth; and if b is less than 1, the function describes exponential decay. Given a table of values for inputs (x) and outputs (y), a linear function would model the relationship if the variables change by a common ratio over given intervals—for every two inputs a given distance apart, the quotients of their corresponding outputs is constant. Here are some sample ordered pairs:

x	0	1	2	3
y	3	6	12	24

For every 1 unit increase in x, the quotient of the corresponding y-values equals:

$$\frac{1}{2} \left(\text{ex.,} \frac{3}{6}, \frac{6}{12}, \frac{12}{24} \right)$$

Therefore, the table represents exponential growth.

Given a scenario describing an exponential function, the growth or decay is expressed using multiplication. Words such as "doubling" and "halving" will often be used. A problem might indicate that the value of an investment triples every year or that every decade the population of an insect is halved. These indicate exponential growth and decay.

The process for writing the equation to represent an exponential model is covered in Section 4.3.

hard to understand ✗

Using Two-Way Tables to Summarize Categorical Data and Relative Frequencies, and Calculate Conditional Probability

Categorical data consists of numerical values found by dividing the entire set into subsets based on variables that represent categories. An example would be the survey results of high school seniors, specifying gender and asking whether they consume alcohol. The data can be arranged in a two-way frequency table (also called a contingency table).

Two-Way Frequency/Contingency Tables   *look left and up for x and y.*

A contingency table presents the frequency tables of both variables simultaneously, as shown below. The levels of one variable constitute the rows of the table, and the levels of the other constitute the columns. The margins consist of the sum of cell frequencies for each row and each column (marginal frequencies). The lower right corner value is the sum of marginal frequencies for the rows or the sum of the marginal frequencies for the columns. Both are equal to the total sample size.

	Drink Alcohol	Do Not Drink Alcohol	Total
Male	63	51	114
Female	37	68	105
Total	100	119	219

Conditional Frequencies

To calculate a conditional relative frequency, the cell frequency is divided by the marginal frequency for the desired outcome given the conditional category. For instance, using the table to determine the relative frequency that a female drinks, the number of females who drink (desired outcome) is divided by the total number of females (conditional category). The conditional relative frequency would equal $\frac{37}{105}$, which equals .35. If a problem asks for a conditional probability, the answer would be expressed as a fraction in simplest form. If asked for a percent, multiply the decimal by 100.

Association of Variables

An association between the variables exists if the conditional relative frequencies are different depending on condition. If the conditional relative frequencies are close to equal, then the variables are independent. For our example, 55% of senior males and 35% of senior females drink alcohol. The difference between frequencies across conditions (male or female) is enough to conclude that an association exists between the variables.

Making Inferences about Population Parameters Based on Sample Data

Statistical inference, based in probability theory, makes calculated assumptions about an entire population based on data from a sample set from that population.

Population Parameters

A population is the entire set of people or things of interest. Suppose a study is intended to determine the number of hours of sleep per night for college females in the U.S. The population would consist of EVERY college female in the country. A sample is a subset of the population that may be used for the study. It would not be practical to survey every female college student, so a sample might consist of 100 students per school from 20 different colleges in the country. From the results of the survey, a sample statistic can be calculated. A sample statistic is a numerical characteristic of the sample data, including mean and variance. A sample statistic can be used to estimate a corresponding population parameter. A population parameter is a numerical characteristic of the entire population. Suppose the sample data had a mean (average) of 5.5. This sample statistic can be used as an estimate of the population parameter (average hours of sleep for every college female in the U.S.).

Confidence Intervals
A population parameter is usually unknown and therefore is estimated using a sample statistic. This estimate may be highly accurate or relatively inaccurate based on errors in sampling. A confidence interval indicates a range of values likely to include the true population parameter. These are constructed at a given confidence level, such as 95%. This means that if the same population is sampled repeatedly, the true population parameter would occur within the interval for 95% of the samples.

Measurement Error
The accuracy of a population parameter based on a sample statistic may also be affected by measurement error, which is the difference between a quantity's true value and its measured value. Measurement error can be divided into random error and systematic error. An example of random error for the previous scenario would be a student reporting 8 hours of sleep when she actually sleeps 7 hours per night. Systematic errors are those attributed to the measurement system. Suppose the sleep survey gave response options of 2, 4, 6, 8, or 10 hours. This would lead to systematic measurement error.

Using Statistics to Investigate Measures of Center of Data and Analyzing Shape, Center, and Spread

Descriptive statistics are used to gain an understanding of properties of a data set. This entails examining the center, spread, and shape of the sample data.

Center
The center of the sample set can be represented by its mean, median, or mode. The mean is the average of the data set, calculated by adding the data values and dividing by the sample size. The median is the value of the data point in the middle when the sample is arranged in numerical order. If the sample has an even number of data points, the mean of the two middle values is the median. The mode is the value that appears most often in a data set. It is possible to have multiple modes (if different values repeat equally as often) or no mode (if no value repeats).

Spread
Methods for determining the spread of the sample include calculating the range and standard deviation for the data. The range is calculated by subtracting the lowest value from the highest value in the set. The standard deviation of the sample can be calculated using the formula:

$$\sigma = \sqrt{\frac{\sum(x - \bar{x})^2}{n - 1}}$$

where $\bar{x}$ = sample mean and n = sample size.

Shape
The shape of the sample when displayed as a histogram or frequency distribution plot helps to determine if the sample is normally distributed (bell-shaped curve), symmetrical, or has measures of skewness (lack of symmetry) or kurtosis. Kurtosis is a measure of whether the data are heavy-tailed (high number of outliers) or light-tailed (low number of outliers).

Evaluating Reports to Make Inferences, Justify Conclusions, and Determine Appropriateness of Data Collection Methods

The presentation of statistics can be manipulated to produce a desired outcome. Here's a statement to consider: "Four out of five dentists recommend our toothpaste." Who are the five dentists? This statement is very different from the statement: "Four out of every five dentists recommend our toothpaste." Whether intentional or unintentional, statistics can be misleading. Statistical reports should be examined to verify the validity and significance of the results. The context of the numerical values allows for deciphering the meaning, intent, and significance of the survey or study. Questions on this material will require students to use critical thinking skills to justify or reject results and conclusions.

When analyzing a report, who conducted the study and their intent should be considered. Was it performed by a neutral party or by a person or group with a vested interest? A study on health risks of smoking performed by a health insurance company would have a much different intent than one performed by a cigarette company. The sampling method and the data collection method should be considered too. Was it a true random sample of the population or was one subgroup over- or underrepresented? The sleep study scenario from Section 8 is one example. If all 20 schools included in the study were state colleges, the results may be biased due to a lack of private school participants. Also, the measurement system used to obtain the data should be noted. Was the system accurate and precise or was it a flawed system? If possible, responses were limited for the sleep study to 2, 4, 6, 8, or 10, it could be argued that the measurement system was flawed.

Every scenario involving statistical reports will be different. The key is to examine all aspects of the study before determining whether to accept or reject the results and corresponding conclusions.

Passport to Advanced Math

Creating a Quadratic or Exponential Function

Quadratic Models
A quadratic function can be written in the standard form: $y = ax^2 + bx + c$. It can be represented by a u-shaped graph called a parabola. For a quadratic function where the value of a is positive, as the inputs increase, the outputs increase until a certain value (maximum of the function) is reached. As inputs increase past the value that corresponds with the maximum output, the relationship reverses and the outputs decrease. For a quadratic function where a is negative, as the inputs increase, the outputs (1) decrease, (2) reach a maximum, and (3) then increase.

Consider a ball thrown straight up into the air. As time passes, the height of the ball increases until it reaches its maximum height. After reaching the maximum height, as time increases, the height of the ball decreases (it is falling toward the ground). This relationship can be expressed as a quadratic function where time is the input (x), and the height of the ball is the output (y).

Given a scenario that can be modeled by a quadratic function, to write its equation, the following is needed: its vertex and any other ordered pair; or any three ordered pairs for the function. Given three ordered pairs, they should be substituted into the general form ($y = ax^2 + bx + c$) to create a system of three equations. For example, given the ordered pairs (2, 3), (3, 13), and (4, 29), it yields:

$$3 = a(2)2 + b(2) + c \rightarrow 4a + 2b + c = 3$$

$$13 = a(3)2 + b(3) + c \rightarrow 9a + 3b + c = 13$$

$$29 = a(4)2 + b(4) + c \rightarrow 16a + 24b + c = 29$$

The values for a, b, and c in the system can be found and substituted into the general form to write the equation of the function. In this case, the equation is:

$$y = 3x^2 - 5x + 1$$

Exponential Models
Exponential functions can be written in the form:

$$y = a \times b^x$$

Scenarios involving growth and decay can be modeled by exponential functions.

The equation for an exponential function can be written given the y-intercept (a) and the growth rate (b). The y-intercept is the output (y) when the input (x) equals zero. It can be thought of as an "original value" or starting point. The value of b is the rate at which the original value increases ($b > 1$) or decreases ($b < 1$). Suppose someone deposits $1200 into a bank account that accrues 1% interest per month. The y-intercept, a, would be $1200, while the growth rate, b, would be 1.01 (100% of the original value + 1% interest). This scenario could be written as the exponential function $y = 1200 \times 1.01^x$, where x represents the number of months since the deposit and y represents money in the account.

Given a scenario that models an exponential function, the equation can also be written when provided two ordered pairs.

Determining the Most Suitable Form of an Expression

It is possible for algebraic expressions and equations to be written that look completely different, yet are still equivalent. For instance, the expression $4(2x - 3) - 3x + 5$ is equivalent to the expression $5x - 7$. Given two algebraic expressions, it can be determined if they are equivalent by writing them in simplest form. Distribution should be used, if applicable, and like terms should be combined. Given two algebraic equations, it can be determined if they are equivalent by solving each for the same variable. Here are two sample equations to consider:

$$3x - 4y = 7$$

$$x + 2 = \frac{4}{3}y + 4\frac{1}{3}$$

kinda obvious but good to know ig.

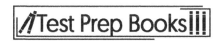

To determine if they are equivalent, solving for x is required.

$$3x - 4y = 7 \qquad\qquad x + 2 = \frac{4}{3}y + 4\frac{1}{3}$$

$$3x = 4y + 7 \qquad\qquad x = \frac{4}{3}y + 2\frac{1}{3}$$

$$x = \frac{4}{3}y + \frac{7}{3} \qquad\qquad x = \frac{4}{3}y + 2\frac{1}{3}$$

The equations are equivalent.

Equivalent Forms of Functions

Equations in two variables can often be written in different forms to easily recognize a given trait of the function or its graph. Linear equations written in slope-intercept form allow for recognition of the slope and y-intercept; and linear equations written in standard form allow for identification of the x and y-intercepts. Quadratic functions written in standard form allow for identification of the y-intercept and for easy calculation of outputs; and quadratic functions written in vertex form allow for identification of the function's minimum or maximum output and its graph's vertex. Polynomial functions written in factored form allow for identification of the zeros of the function.

The method of substituting the same inputs (x-values) into functions to determine if they produce the same outputs can reveal if functions are not equivalent (different outputs). However, corresponding inputs and outputs do not necessarily indicate equivalent functions.

Create Equivalent Expressions Involving Rational Exponents

Converting To and From Radical Form

Algebraic expressions involving radicals ($\sqrt{}$, $\sqrt[3]{}$, etc.) can be written without the radical by using rational (fraction) exponents. For radical expressions, the value under the root symbol is called the radicand, and the type of root determines the index. For example, the expression $\sqrt{6x}$ has a radicand of 6x and index of 2 (it is a square root). If the exponent of the radicand is 1, then $\sqrt[n]{a} = a^{\frac{1}{n}}$ where n is the index. A number or variable without a power has an implied exponent of 1. For example:

$$\sqrt{6} = 6^{\frac{1}{2}}$$

And

$$125^{\frac{1}{3}} = \sqrt[3]{125}$$

For any exponent of the radicand:

$$\sqrt[n]{a^m} = \left(\sqrt[n]{a}\right)^m = a^{\frac{m}{n}}$$

For example:

$$64^{\frac{5}{3}} = \sqrt[3]{64^5} \, or \left(\sqrt[3]{64}\right)^5$$

and:

$$(xy)^{\frac{2}{3}} = \sqrt[3]{(xy)^2} \, or \left(\sqrt[3]{xy}\right)^2$$

Simplifying Expressions with Rational Exponents

When simplifying expressions with rational exponents, all basic properties for exponents hold true. When multiplying powers of the same base (same value with or without the same exponent), the exponents are added. For example:

$$x^{\frac{2}{7}} \times x^{\frac{3}{14}} = x^{\frac{1}{2}} \left(\frac{2}{7} + \frac{3}{14} = \frac{1}{2}\right)$$

When dividing powers of the same base, the exponents are subtracted. For example:

$$\frac{5^{\frac{2}{3}}}{5^{\frac{1}{2}}} = 5^{\frac{1}{6}} \left(\frac{2}{3} - \frac{1}{2} = \frac{1}{6}\right)$$

When raising a power to a power, the exponents are multiplied. For example:

$$\left(5^{\frac{1}{2}}\right)^4 = 5^2 \left(\frac{1}{2} \times 4 = 2\right)$$

When simplifying expressions with exponents, a number should never be raised to a power or a negative exponent. If a number has an integer exponent, its value should be determined. If the number has a rational exponent, it should be rewritten as a radical and the value determined if possible. A base with a negative exponent moves from the numerator to the denominator of a fraction (or vice versa) and is written with a positive exponent. For example:

$$x^{-3} = \frac{1}{x^3}$$

and

$$\frac{2}{5x^{-2}} = \frac{2x^2}{5}$$

The exponent of 5 is 1, and therefore the 5 does not move.

Here's a sample expression:

$$(27x^{-9})^{\frac{1}{3}}$$

After the implied exponents are noted, a power should be raised to a power by multiplying exponents, which yields $27^{\frac{1}{3}}x^{-3}$. Next, the negative exponent is eliminated by moving the base and power:

$$\frac{27^{\frac{1}{3}}}{x^3}$$

Then the value of the number is determined to a power by writing it in radical form:

$$\frac{\sqrt[3]{27}}{x^3}$$

Simplifying yields:

$$\frac{3}{x^3}$$

Creating an Equivalent Form of an Algebraic Expression

There are many different ways to write algebraic expressions and equations that are equivalent to each other. Converting expressions from standard form to factored form and vice versa are skills commonly used in advanced mathematics. Standard form of an expression arranges terms with variables powers in descending order (highest exponent to lowest and then constants). Factored form displays an expression as the product of its factors (what can be multiplied to produce the expression).

Converting Standard Form to Factored Form

To factor an expression, a greatest common factor needs to be factored out first. Then, if possible, the remaining expression needs to be factored into the product of binomials. A binomial is an expression with two terms.

Greatest Common Factor

The greatest common factor (GCF) of a monomial (one term) consists of the largest number that divides evenly into all coefficients (number part of a term); and if all terms contain the same variable, the variable with the lowest exponent. The GCF of $3x^4 - 9x^3 + 12x^2$ would be $3x^2$. To write the factored expression, every term needs to be divided by the GCF, then the product of the resulting quotient and the GCF (using parentheses to show multiplication) should be written. For the previous example, the factored expression would be:

$$3x^2(x^2 - 3x + 4)$$

Factoring Ax² + Bx + C When A = 1

To factor a quadratic expression in standard form when the value of a is equal to 1, the factors that multiply to equal the value of c should be found and then added to equal the value of b (the signs of b and c should be included). The factored form for the expression will be the product of binomials: $(x + $

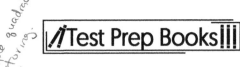

$factor1)(x + factor2)$. Here's a sample expression: $x^2 - 4x - 5$. The two factors that multiply to equal c(-5) and add together to equal b(-4) are -5 and 1. Therefore, the factored expression would be $(x - 5)(x + 1)$. Note $(x + 1)(x - 5)$ is equivalent.

Factoring a Difference of Squares

A difference of squares is a binomial expression where both terms are perfect squares (perfect square-perfect square). Perfect squares include 1, 4, 9, 16 . . . and x^2, x^4, x^6 . . .

The factored form of a difference of squares will be:

$$(\sqrt{term1} + \sqrt{term2})(\sqrt{term1} - \sqrt{term2})$$

For example:

$$x^2 - 4 = (x + 2)(x - 2)$$

And

$$25x^6 - 81 = (5x^3 + 9)(5x^3 - 9)$$

Factoring $Ax^2 + Bx + C$ when $A \neq 1$

To factor a quadratic expression in standard form when the value of a is not equal to 1, the factors that multiply to equal the value of $a \times c$ should be found and then added to equal the value of b. Next, the expression splitting the bx term should be rewritten using those factors. Instead of three terms, there will now be four. Then the first two terms should be factored using GCF, and a common binomial should be factored from the last two terms. The factored form will be: (common binomial) (2 terms out of binomials). In the sample expression $2 \times 2 + 11x + 12$, the value of $a \times c$ ($2x12$) = 24. Two factors that multiply to 24 and added together to yield b(11) are 8 and 3. The bx term (11x) can be rewritten by splitting it into the factors:

$$2 \times 2 + 8x + 3x + 12$$

A GCF from the first two terms can be factored as:

$$2x(x + 4) + 3x + 12$$

A common binomial from the last two terms can then be factored as:

$$2(x + 4) + 3(x + 4)$$

The factored form can be written as a product of binomials:

$$(x + 4)(2x + 3)$$

Converting Factored Form to Standard Form

To convert an expression from factored form to standard form, the factors are multiplied.

Solving a Quadratic Equation

A quadratic equation is one in which the highest exponent of the variable is 2. A quadratic equation can have two, one, or zero real solutions. Depending on its structure, a quadratic equation can be solved by (1) factoring, (2) taking square roots, or (3) using the quadratic formula.

Solving Quadratic Equations by Factoring

To solve a quadratic equation by factoring, the equation should first be manipulated to set the quadratic expression equal to zero. Next, the quadratic expression should be factored using the appropriate method(s). Then each factor should be set equal to zero. If two factors multiply to equal zero, then one or both factors must equal zero. Finally, each equation should be solved. Here's a sample

$$x^2 - 10 = 3x - 6$$

The expression should be set equal to zero:

$$x^2 - 3x - 4 = 0$$

The expression should be factored:

$$(x - 4)(x + 1) = 0$$

Each factor should be set equal to zero: $x - 4 = 0; x + 1 = 0$. Solving yields $x = 4$ or $x = -1$.

Solving Quadratic Equations by Taking Square Roots

If a quadratic equation does not have a linear term (variable to the first power), it can be solved by taking square roots. This means x^2 needs to be isolated and then the square root of both sides of the equation should be isolated. There will be two solutions because square roots can be positive or negative. ($\sqrt{4}$ = 2 or -2 because $2 \times 2 = 4$ and $-2 \times -2 = 4$.) Here's a sample equation:

$$3x^2 - 12 = 0$$

Isolating x^2 yields $x^2 = 4$. The square root of both sides is then solved: $x = 2$ or -2.

The Quadratic Formula

When a quadratic expression cannot be factored or is difficult to factor, the quadratic formula can be used to solve the equation. To do so, the equation must be in the form:

$$ax^2 + bx + c = 0$$

The quadratic formula is:

$$x = \frac{-b \pm \sqrt{b^2 - 4ac}}{2a}$$

(The $\pm$ symbol indicates that two calculations are necessary, one using $+$ and one using $-$.) Here's a sample equation:

$$3x^2 - 2x = 3x + 2$$

First, the quadratic expression should be set equal to zero:

$$3x^2 - 5x - 2 = 0$$

Then the values are substituted for *a*(3), *b*(-5), and *c*(-2) into the formula:

$$x = \frac{-(-5) \pm \sqrt{(-5)^2 - 4(3)(-2)}}{2(3)}$$

Simplification yields:

$$x = \frac{5 \pm \sqrt{49}}{6} \rightarrow x = \frac{5 \pm 7}{6}$$

Calculating two values for x using $+$ and $-$ yields:

$$x = \frac{5 + 7}{6}; x = \frac{5 - 7}{6}$$

Simplification yields:

$$x = 2 \text{ or } -\frac{1}{3}.$$

Just as with any equation, solutions should be checked by substituting the value into the original equation.

Adding, Subtracting, and Multiplying Polynomial Expressions

A polynomial expression is a monomial (one term) or the sum of monomials (more than one term separated by addition or subtraction). A polynomial in standard form consists of terms with variables written in descending exponential order and with any like terms combined.

Adding/Subtracting Polynomials

When adding or subtracting polynomials, each polynomial should be written in parenthesis; the negative sign should be distributed when necessary, and like terms need to be combined. Here's a sample equation: add $3x^3 + 4x - 3$ to $x^3 - 3x^2 + 2x - 2$. The sum is set as follows:

$$(x^3 - 3x^3 + 2x - 2) + (3x^3 + 4x - 3)$$

In front of each set of parentheses is an implied positive 1, which, when distributed, does not change any of the terms. Therefore, the parentheses should be dropped and like terms should be combined:

$$x^3 - 3x^2 + 2x - 2 + 3x^3 + 4x - 3$$

$$4x^3 - 3x^2 + 6x - 5$$

Here's another sample equation: subtract $3x^3 + 4x - 3$ from $x^3 - 3x^2 + 2x - 2$. The difference should be set as follows:

$$(x^3 - 3x^2 + 2x - 2) - (3x^3 + 4x - 3)$$

The implied $+1$ in front of the first set of parentheses will not change those four terms; however, distributing the implied -1 in front of the second set of parentheses will change the sign of each of those three terms:

$$x^3 - 3x^2 + 2x - 2 - 3x^3 - 4x + 3$$

Combining like terms yields:

$$-2x^3 - 3x^2 - 2x + 1$$

Multiplying Polynomials

When multiplying monomials, the coefficients are multiplied and exponents of the same variable are added. For example:

$$-5x^3y^2z \times 2x^2y^5z^3 = -10x^5y^7z^4$$

When multiplying polynomials, the monomials should be distributed and multiplied, then any like terms should be combined and written in standard form. Here's a sample equation:

$$2x^3(3x^2 + 2x - 4)$$

First, $2x^3$ should be multiplied by each of the three terms in parentheses:

$$2x^3 \times 3x^2 + 2x^3 \times 2x + 2x^3 \times -4$$

$$6x^5 + 4x^4 - 8x^3$$

Multiplying binomials will sometimes be taught using the FOIL method (where the products of the first, outside, inside, and last terms are added together). However, it may be easier and more consistent to think of it in terms of distributing. Both terms of the first binomial should be distributed to both terms of the second binomial. For example, the product of binomials $(2x + 3)(x - 4)$ can be calculated by distributing $2x$ and distributing 3:

$$2x \times x + 2x \times -4 + 3 \times x + 3 \times -4$$

$$2x^2 - 8x + 3x - 12$$

Combining like terms yields $2x^2 - 5x - 12$.

The general principle of distributing each term can be applied when multiplying polynomials of any size. To multiply

$$(x^2 + 3x - 1)(5x^3 - 2x^2 + 2x + 3)$$

all three terms should be distributed from the first polynomial to each of the four terms in the second polynomial and then any like terms should be combined. If a problem requires multiplying more than two polynomials, two at a time can be multiplied and combined until all have been multiplied. To multiply $(x + 3)(2x - 1)(x + 5)$, two polynomials should be chosen to multiply together first. Multiplying the last two results in:

$$(2x - 1)(x + 5) = 2x^2 + 9x - 5$$

That product should then be multiplied by the third polynomial:

$$(x + 3)(2x^2 + 9x - 5)$$

The final answer should equal:

$$2x^3 + 15x^2 + 36x - 15.$$

Solving an Equation in One Variable that Contains Radicals or Contains the Variable in the Denominator of a Fraction

Equations with radicals containing numbers only as the radicand are solved the same way that an equation without a radical would be. For example, $3x + \sqrt{81} = 45$ would be solved using the same steps as if solving $2x + 4 = 12$. Radical equations are those in which the variable is part of the radicand. For example, $\sqrt{5x + 1} - 6 = 0$ and $\sqrt{x - 3} + 5 = x$ would be considered radical equations.

Radical Equations

To solve a radical equation, the radical should be isolated and both sides of the equation should be raised to the same power to cancel the radical. Raising both sides to the second power will cancel a square root, raising to the third power will cancel a cube root, etc. To solve $\sqrt{5x + 1} - 6 = 0$, the radical should be isolated first: $\sqrt{5x + 1} = 6$. Then both sides should be raised to the second power:

$$(\sqrt{5x + 1})^2 = (6)^2 \rightarrow 5x + 1 = 36$$

Lastly, the linear equation should be solved: $x = 7$.

Radical Equations with Extraneous Solutions

If a radical equation contains a variable in the radicand and a variable outside of the radicand, it must be checked for extraneous solutions. An extraneous solution is one obtained by following the proper process for solving an equation but does not "check out" when substituted into the original equation. Here's a sample equation:

$$\sqrt{x - 3} + 5 = x$$

Isolating the radical yields:

$$\sqrt{x - 3} = x - 5$$

Next, both sides should be squared to cancel the radical:

$$\left(\sqrt{x - 3}\right)^2$$

$$(x - 5)^2 \rightarrow x - 3$$

$$(x - 5)(x - 5)$$

The binomials should be multiplied: $x - 3 = x^2 - 10x + 25$. The quadratic equation is then solved:

$$0 = x^2 - 11x + 28$$

$$0 = (x - 7)(x - 4)$$

$$x - 7 = 0; x - 4 = 0$$

$$x = 7 \text{ or } x = 4$$

To check for extraneous solutions, each answer can be substituted, one at a time, into the original equation. Substituting 7 for x, results in $7 = 7$. Therefore, 7 is a solution. Substituting 4 for x results in $6 = 4$. This is false; therefore, 4 is an extraneous solution.

Equations with a Variable in the Denominator of a Fraction

For equations with variables in the denominator, if the equation contains two rational expressions (on opposite sides of the equation, or on the same side and equal to zero), it can be solved like a proportion. Here's an equation to consider:

$$\frac{5}{2x-2} = \frac{15}{x^2-1}$$

First, cross-multiplying yields:

$$5(x^2-1) = 15(2x-2)$$

Distributing yields:

$$5x^2 - 5 = 30x - 30$$

In solving the quadratic equation (see section *Solving a Quadratic Equation*), it is determined that $x = 1$ or $x = 5$. Solutions must be checked to see if they are extraneous. Extraneous solutions either produce a false statement when substituted into the original equation or create a rational expression with a denominator of zero (dividing by zero is undefined). Substituting 5 into the original equation produces $\frac{5}{8} = \frac{5}{8}$; therefore, 5 is a solution. Substituting 1 into the original equation results in both denominators equal to zero; therefore, 1 is an extraneous solution.

If an equation contains three or more rational expressions: the least common denominator (LCD) needs to be found for all the expressions, then both sides of the equation should be multiplied by the LCD. The LCD consists of the lowest number that all coefficients divide evenly into and for every variable, the highest power of that variable. Here's a sample equation:

$$\frac{3}{5x} - \frac{4}{3x} = \frac{1}{3}$$

The LCD would be $15x$. Both sides of the equation should be multiplied by $15x$:

$$15x\left(\frac{3}{5x} - \frac{4}{3x}\right) = 15x\left(\frac{1}{3}\right)$$

$$\frac{45x}{5x} - \frac{60x}{3x} = \frac{15x}{3}$$

$$9 - 20 = 5x$$

$$x = -2\frac{1}{2}$$

Any extraneous solutions should be identified.

Solving a System of One Linear Equation and One Quadratic Equation

A system of equations consists of two variables in two equations. A solution to the system is an ordered pair (x, y) that makes both equations true. When displayed graphically, a solution to a system is a point of intersection between the graphs of the equations. When a system consists of one linear equation and one quadratic equation, there may be one, two, or no solutions. If the line and parabola intersect at two points, there are two solutions to the system; if they intersect at one point, there is one solution; if they do not intersect, there is no solution.

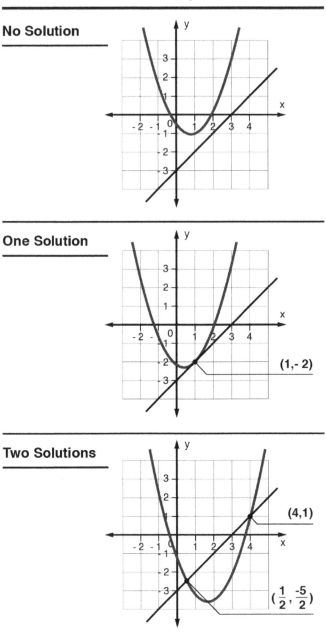

Systems with One Linear Equation and One Quadratic Equation

No Solution

One Solution

(1,- 2)

Two Solutions

(4,1)

$(\frac{1}{2}, \frac{-5}{2})$

One method for solving a system of one linear equation and one quadratic equation is to graph both functions and identify point(s) of intersection. This, however, is not always practical. Graph paper may not be available or the intersection points may not be easily identified. Solving the system algebraically involves using the substitution method.

Consider the following system:

$$y = x^2 + 9x + 11; y = 2x - 1$$

The equivalent value of y should be substituted from the linear equation $(2x - 1)$ into the quadratic equation. The resulting equation is:

$$2x - 1 = x^2 + 9x + 11$$

Next, this quadratic equation should be solved using the appropriate method: factoring, taking square roots, or using the quadratic formula (see section *Solving a Quadratic Equation*). Solving this quadratic equation by factoring results in $x = -4$ or $x = -3$. Next, the corresponding y-values should be found by substituting the x-values into the original linear equation:

$$y = 2(-4) - 1; y = 2(-3) - 1$$

The solutions should be written as ordered pairs: (-4, -9) and (-3, -7). Finally, the possible solutions should be checked by substituting each into both of the original equations. In this case, both solutions "check out."

Rewriting Simple Rational Expressions

A rational expression is an algebraic expression including variables that look like a fraction. In simplest form, the numerator and denominator of a rational expression do not have common divisors (factors). To simplify a rational expression, the numerator and denominator (see Section 4) should be factored; then any common factors in the numerator and denominator should be canceled. To simplify the foll, the numerator and denominator should be written as a product of its factors:

$$\frac{3x^2y}{12xy^3}$$

$$\frac{3 \cdot x \cdot x \cdot y}{2 \cdot 2 \cdot 3 \cdot x \cdot y \cdot y \cdot y}$$

Canceling common factors leaves: $\frac{x}{2 \cdot 2 \cdot y \cdot y}$. Multiplying the remaining factors results in $\frac{x}{4y^2}$.

Here's a rational expression:

$$\frac{x^2 - 1}{x^2 - x - 2}$$

Factoring the numerator and denominator produces:

$$\frac{(x + 1)(x - 1)}{(X - 2)(x + 1)}$$

Test Prep Books!!!

Each binomial in parentheses is a factor and only the exact same binomial would cancel that factor. By canceling factors, the expression is simplified to:

$$\frac{x-1}{x-2}$$

The variable x itself is not a factor. Therefore, they do not cancel each other out.

Multiplying/Dividing Rational Expressions

When multiplying or dividing rational expressions, the basic concepts of operations with fractions are used. To multiply, (1) all numerators and denominators need to be factored, (2) common factors should be canceled between any numerator and any denominator, (3) the remaining factors of the numerator and the remaining factors of the denominator should be multiplied, and (4) the expression should be checked to see whether it can be simplified further.

Steps

To multiply the following, each numerator and denominator should be written as a product of its factors:

$$\frac{4a^4}{3} \times \frac{6}{5a^2}$$

$$\frac{2 \cdot 2 \cdot a \cdot a \cdot a \cdot a}{3} \times \frac{3 \cdot 2}{5 \cdot a \cdot a}$$

After canceling common factors, the remaining expression is:

$$\frac{2 \cdot 2 \cdot a \cdot a}{1} \times \frac{2}{5}$$

A factor of 1 remains if all others are canceled. Multiplying remaining factors produces:

$$\frac{8a^2}{5}$$

To divide rational expressions, the expression should be changed to multiplying by the reciprocal of the divisor (just as with fractions: $\frac{1}{2} \div \frac{3}{4} = \frac{1}{2} \times \frac{4}{3}$); then follow the process for multiplying rational expressions.

Here's a sample expression:

$$\frac{2x}{x^2-16} \div \frac{4x^2+6x}{x^2+6x+8}$$

First, the division problem should be changed to a multiplication problem:

$$\frac{2x}{x^2-16} \times \frac{x^2+6x+8}{4x^2+6x}$$

Then, the equation should be factored:

$$\frac{2x}{(x+4)(x-4)} \times \frac{(x+4)(x+2)}{2x(2x+3)}$$

104

Canceling yields:

$$\frac{1}{(x-4)} \times \frac{(x+2)}{(2x+3)}$$

Multiplying the remaining factors produces:

$$\frac{x+2}{2x^2 - 5x - 12}$$

Adding/Subtracting Rational Expressions

Just as with adding and subtracting fractions, to add or subtract rational expressions, a common denominator is needed. (The numerator is added or subtracted and the denominator stays the same.) If the expressions have like denominators, subtraction should be changed to add the opposite (a -1 is distributed to each term in the numerator of the expression being subtracted); the denominators should be factored and the expressions added; the numerator should then be factored; and the equation should be simplified if possible. Here's a sample expression:

$$\frac{2x^2 + 4x - 3}{x+3} - \frac{x^2 - 2x - 12}{x+3}$$

Changing subtraction to add the opposite yields:

$$\frac{2x^2 + 4x - 3}{x+3} + \frac{-x^2 + 2x + 12}{x+3}$$

The denominator cannot be factored, so the expression should be added, resulting in:

$$\frac{x^2 + 6x + 9}{x+3}$$

Simplification is performed by factoring the numerator:

$$\frac{(x+3)(x+3)}{(x+3)}$$

Canceling yields: $\frac{x+3}{1}$, or simply x + 3.

To add or subtract rational expressions with unlike denominators, the denominators must be changed by finding the least common multiple (LCM) of the expressions. To find the LCM, each expression should be factored and the product should be formed using each factor the greatest number of times it occurs. The LCM of $12xy^2$ and $15x^3y$ would be $60x^3y^2$. The LCM of $x^2 + 5x + 4$ (which factors to $(x+4)(x+1)$) and $x^2 + 2x + 1$ (which factors to $(x+1)(x+1)$) would be $(x+4)(x+1)(x+1)$.

To add or subtract expressions with unlike denominators: (1) subtraction should be changed to add the opposite; (2) the denominators are factored; (3) an LCM should be determined for the denominators; (4) the numerator and denominator of each expression should be multiplied by the missing factor(s); (5) the expressions that now have like denominators should be added; (6) the numerator should be factored; and (7) simplification should be performed if possible.

$\left.\right\}$ steps.

Here's a sample expression:

$$\frac{x^2 + 6x + 11}{x^2 + 7x + 12} - \frac{2}{x + 3}$$

First, subtraction should be changed to addition:

$$\frac{x^2 + 6x + 11}{x^2 + 7x + 12} + \frac{-2}{x + 3}$$

Then, the denominators are factored:

$$\frac{x^2 + 6x + 11}{(x + 4)(x + 3)} + \frac{-2}{x + 3}$$

The LCM of $(x + 4)(x + 3)$ and $(x + 3)$ should be determined, which is $(x + 4)(x + 3)$. The numerator and denominator should be multiplied by the missing factor:

$$\frac{x^2 + 6x + 11}{(x + 4)(x + 3)} + \frac{-2}{x + 3} \times \frac{(x + 4)}{(x + 4)} = \frac{x^2 + 6x + 11}{(x + 4)(x + 3)} + \frac{-2x - 8}{(x + 4)(x + 3)}$$

The expressions should be added, resulting in:

$$\frac{x^2 + 4x + 3}{(x + 4)(x + 3)}$$

The numerator should be factored:

$$\frac{(x + 3)(x + 1)}{(x + 4)(x + 3)}$$

Simplifying yields:

$$\frac{x + 1}{x + 4}$$

Interpreting Parts of Nonlinear Expressions in Terms of Their Context

When a nonlinear function is used to model a real-life scenario, some aspects of the function may be relevant while others may not. The context of each scenario will dictate what should be used. In general, x- and y-intercepts will be points of interest. A y-intercept is the value of y when x = zero; and an x-intercept is the value of x when y = zero. Suppose a nonlinear function models the value of an investment(y) over the course of time(x). It would be relevant to determine the initial value (the y-intercept where time = zero), as well as any point in time in which the value would be zero (the x-intercept).

Another aspect of a function that is typically desired is the rate of change. This tells how fast the outputs are growing or decaying with respect to given inputs. For more on rates of change regarding quadratic and exponential functions, see Sections 5 and 6. For polynomial functions, the rate of change can be estimated by the highest power of the function. Polynomial functions also include absolute and/or

relative minimums and maximums. Functions modeling production or expenses should be considered. Maximum and minimum values would be relevant aspects of these models.

Finally, the domain and range for a function should be considered for relevance. The domain consists of all input values and the range consists of all output values. For instance, a function could model the volume of a container to be produced in relation to its height. Although the function that models the scenario may include negative values for inputs and outputs, these parts of the function would obviously not be relevant.

Understanding the Relationship Between Zeros and Factors of Polynomials

The zeros of a function are the x-intercepts of its graph. They are called zeros because they are the x-values for which $y = 0$.

Finding Zeros
To find the zeros of a polynomial function, it should be written in factored form (see Section 4), then each factor should be set equal to zero and solved. To find the zeros of the function $y = 3x^3 - 3x^2 - 36x$, the polynomial should be factored first. Factoring out a GCF results in:

$$y = 3x(x^2 - x - 12)$$

Then factoring the quadratic function yields:

$$y = 3x(x - 4)(x + 3)$$

Next, each factor should be set equal to zero: $3x = 0; x - 4 = 0; x + 3 = 0$. By solving each equation, it is determined that the function has zeros, or x-intercepts, at 0, 4, and -3.

Writing a Polynomial with Given Zeros
Given zeros for a polynomial function, to write the function, a linear factor corresponding to each zero should be written. The linear factor will be the opposite value of the zero added to x. Then the factors should be multiplied and the function written in standard form (see Section 6). To write a polynomial with zeros at -2, 3, and 3, three linear factors should be written:

obviously:
x = 3;
x - 3 = 0

$$y = (x + 2)(x - 3)(x - 3)$$

Then, multiplication is used to convert the equation to standard form, producing:

$$y = x^3 - 4x^2 - 3x + 18$$

Dividing Polynomials by Linear Factors
To determine if a linear binomial is a factor of a polynomial, the polynomial should be divided by the binomial. If there is no remainder (it divides evenly), then the binomial is a factor of the polynomial. To determine if a value is a zero of a function, a binomial can be written from that zero and tested by division. To divide a polynomial by a linear factor, the terms of the dividend should be divided by the linear term of the divisor; the same process as long division of numbers (divide, multiply, subtract, drop down, and repeat) should be followed.

$$divisor\sqrt{quotient} \over dividend$$

Remember that when subtracting a binomial, the signs of both terms should be changed. Here's a sample equation: divide $9x^3 - 18x^2 - x + 2$ by $3x + 1$. First, the problem should be set up as long division:

$$3x + 1 \overline{)\; 9x^3 - 18x^2 - x + 2}$$

Then the first term of the dividend ($9x^3$) should be divided by the linear term of the divisor ($3x$):

$$
\begin{array}{r}
3x^2 \\
3x + 1 \overline{)\; 9x^3 - 18x^2 - x + 2}
\end{array}
$$

Next, the divisor should be multiplied by that term of the quotient:

$$
\begin{array}{r}
3x^2 - 7x + 2 \\
3x + 1 \overline{)\; 9x^3 - 18x^2 - x + 2} \\
-9x^3 - 3x^2
\end{array}
$$

Subtraction should come next:

$$
\begin{array}{r}
3x^2 - 7x + 2 \\
3x + 1 \overline{)\; 9x^3 - 18x^2 - x + 2} \\
\underline{-9x^3 - 3x^2 } \\
-21x^2
\end{array}
$$

Now, the next term (-x) should be dropped down:

$$
\begin{array}{r}
3x^2 - 7x + 2 \\
3x + 1 \overline{)\; 9x^3 - 18x^2 - x + 2} \\
\underline{-9x^3 - 3x^2 } \\
-21x^2 - x
\end{array}
$$

Then the process should be repeated, dividing -21x^2 by 3x:

$$
\begin{array}{r}
3x^2 - 7x + 2 \\
3x + 1 \overline{)\; 9x^3 - 18x^2 - x + 2} \\
\underline{-9x^3 - 3x^2 } \\
-21x^2 - x \\
\underline{+21x^2 + 7x } \\
6x
\end{array}
$$

The next term (2) should be dropped and repeated by dividing $6x$ by $3x$:

$$
\require{enclose}
\begin{array}{r}
3x^2 - 7x + 2 \\[-3pt]
3x + 1 \enclose{longdiv}{\;9x^3 - 18x^2 - \;\;x + 2} \\
\underline{-9x^3 - \;\;3x^2} \\
-21x^2 - \;\;x \\
\underline{+21x^2 + 7x} \\
6x + 2 \\
\underline{-6x - 2} \\
0
\end{array}
$$

There is no remainder; therefore, $3x + 1$ is a factor of:

$$9x^3 - 18x^2 - x + 2$$

By the definition of factors:

$$(3x + 1)(3x^2 - 7x + 2) = 9x^3 - 18x^2 - x + 2$$

The quadratic expression can further be factored to produce:

$$(3x + 1)(3x - 1)(x - 2)$$

Understanding a Nonlinear Relationship Between Two Variables

Questions on this material will assess the ability of test takers to make connections between linear or nonlinear equations and their graphical representations. It will also require interpreting graphs in relation to systems of equations. Graphical representations of linear, quadratic, and exponential functions are covered in Section 9 of the algebra content and Sections 4 and 5 of the data analysis content. Graphical representations of systems of equations are covered in Section 4 of the algebra content and Section 8 of the advanced math content.

Graphs of Polynomial Functions

A polynomial function consists of a monomial or sum of monomials arranged in descending exponential order. The graph of a polynomial function is a smooth continuous curve that extends infinitely on both ends. From the equation of a polynomial function, the following can be determined: (1) the end behavior of the graph—does it rise or fall to the left and to the right; (2) the y-intercept and x-intercept(s) and whether the graph simply touches or passes through each x-intercept; and (3) the largest possible number of turning points, where the curve changes from rising to falling or vice versa. To graph the function, these three aspects of the graph should be determined and extra points between the intercepts should be found if necessary.

things I should be able to find from a polynomial for the test.

End Behavior

The end behavior of the graph of a polynomial function can be determined by the degree of the function (largest exponent) and the leading coefficient (coefficient of the term with the largest exponent). There are four possible scenarios for the end behavior: (1) if the degree is odd and the coefficient is positive, the graph falls to the left and rises to the right; (2) if the degree is odd and the coefficient is negative,

AKA leading coefficient rule.

the graph rises to the left and falls to the right; (3) if the degree is even and the coefficient is positive, the graph rises to the left and rises to the right, or (4) if the degree is even and the coefficient is negative, the graph falls to the left and falls to the right.

Take a look at this chart:

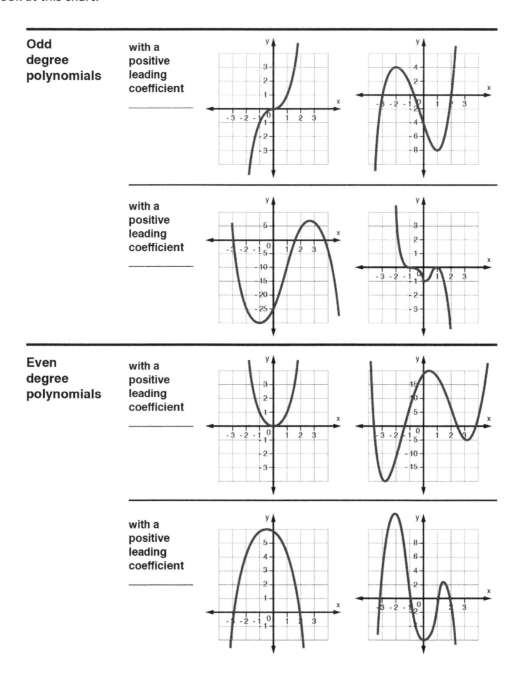

X and Y-Intercepts

The *y*-intercept for any function is the point at which the graph crosses the y-axis. At this point $x = 0$; therefore, to determine the *y*-intercept, $x = 0$ should be substituted into the function and solved for *y*. Finding *x*-intercepts, also called zeros, is covered in Section 11. For a given zero of a function, the graph can either pass through that point or simply touch that point (the graph turns at that zero). This is

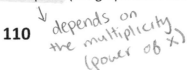
↓ depends on the multiplicity (power of x)

determined by the multiplicity of that zero. The multiplicity of a zero is the number of times its corresponding factor is multiplied to obtain the function in standard form.

For example:

$$y = x^3 - 4x^2 - 3x + 18$$

can be written in factored form as:

$$y = (x + 2)(x - 3)(x - 3) \text{ or } y = (x + 2)(x - 3)^2$$

The zeros of the function would be -2 and 3. The zero at -2 would have a multiplicity of 1, and the zero ~~(bounces)~~ at 3 would have a multiplicity of 2. If a zero has an even multiplicity, then the graph touches the *x*-axis at that zero and turns around. If a zero has an odd multiplicity, then the graph crosses the *x*-axis at that zero.

Turning Points
The graph of a polynomial function can have, at most, a number of turning points equal to one less than the degree of the function. It is possible to have fewer turning points than this value. For example, the function $y = 3x^5 + 2x^2 - 3x$ could have no more than four turning points.

Using Function Notation, and Interpreting Statements Using Function Notation.

Function notation is covered in the *Function/Linear Equation Notation* section under *Heart of Algebra*.

Addition, Subtraction, Multiplication and Division of Functions
Functions denoted by *f*(*x*), *g*(*x*), etc., can be added, subtracted, multiplied, or divided. For example, the function $f(x) = 15x + 100$ represents the cost to have a catered party at a banquet hall (where *x* represents the number of guests); and the function $g(x) = 10x$ represents the cost for unlimited drinks at the party. The total cost of a catered party with unlimited drinks can be represented by adding the functions *f*(*x*) and *g*(*x*). In this case:

$$f(x) + g(x) = (15x + 100) + (10x)$$

Therefore:

$$f(x) + g(x) = 25x + 100$$

$(f(x) + g(x)$ can also be written $(f + g)(x))$. To add, subtract, multiply, or divide functions, the values of the functions should be substituted and the rules for operations with polynomials should be followed (see Sections 6 and 9). It should be noted:

$$(f - g)(x) = f(x) - g(x); (f \times g)(x) = f(x) \times g(x)$$

and

$$\left(\frac{f}{g}\right)(x) = \frac{f(x)}{g(x)}$$

111

Composition of Functions

A composite function is one in which two functions are combined such that the output from the first function becomes the input for the second function (one function should be applied after another function). The composition of a function written as $(g \circ f)(x)$ or $g(f(x))$ is read "g of f of x." The inner function, $f(x)$, would be evaluated first and the answer would be used as the input of the outer function, $g(x)$. To determine the value of a composite function, the value of the inner function should be substituted for the variable of the outer function.

Here's a sample problem:

A store is offering a 20% discount on all of its merchandise. You have a coupon worth $5 off any item.

The cost of an item with the 20% discount can be modeled by the function: $d(x) = 0.8x$. The cost of an item with the coupon can be modeled by the function $c(x) = x - 5$. A composition of functions to model the cost of an item applying the discount first and then the coupon would be $c(d(x))$. Replacing $d(x)$ with its value $(0.8x)$ results in $c(0.8x)$. By evaluating the function $c(x)$ with an input of $0.8x$, it is determined that:

$$c\big(d(x)\big) = 0.8x - 5$$

To model the cost of an item if the coupon is applied first and then the discount, $d(c(x))$ should be determined. The result would be:

$$d(c(x) = 0.8x - 4$$

Evaluating Functions

If a problem asks to evaluate with operations between functions, the new function should be determined and then the given value should be substituted as the input of the new function. To find $(f \times g)(3)$ given $f(x) = x + 1$ and $g(x) = 2x - 3$, the following should be determined:

$$(f \times g)(x)$$

$$f(x) \times g(x)$$

$$(x + 1)(2x - 3)$$

$$2x^2 - x - 3$$

Therefore, $(f \times g)(x) = 2x^2 - x - 3$. To find $(f \times g)(3)$, the function $(f \times g)(x)$ needs to be evaluated for an input of 3:

$$(f \times g)(3)$$

$$2(3)^2 - (3) - 3 = 12$$

or $(f \times g)(3)$

$(3+1)(6-3)$

$(4)(3)$

$\boxed{12}$

Therefore:

$$(f \times g)(3) = 12$$

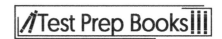

Using Structure to Isolate or Identify a Quantity of Interest

Formulas are mathematical expressions that define the value of one quantity given the value of one or more different quantities. A formula or equation expressed in terms of one variable can be manipulated to express the relationship in terms of any other variable. The equation $y = 3x + 2$ is expressed in terms of the variable y. By manipulating the equation, it can be written as $x = \frac{y-2}{3}$, which is expressed in terms of the variable x. To manipulate an equation or formula to solve for a variable of interest, how the equation would be solved if all other variables were numbers should be considered. The same steps for solving should be followed, leaving operations in terms of the variables, instead of calculating numerical values.

The formula $P = 2l + 2w$ expresses how to calculate the perimeter of a rectangle given its length and width. To write a formula to calculate the width of a rectangle given its length and perimeter, the previous formula relating the three variables should be used and the variable w should be solved. If P and l were numerical values, this would be a two-step linear equation solved by subtraction and division. To solve the equation $P = 2l + 2w$ for w, $2l$ should be subtracted from both sides: $P - 2l = 2w$. Then both sides should be divided by 2:

$$\frac{P - 2l}{2} = w \text{ or } \frac{P}{2} - l = w$$

The distance formula between two points on a coordinate plane can be found using the formula:

$$d = \sqrt{(x_2 - x_1)^2 + (y_2 - y_1)^2}$$

A problem might require determining the x-coordinate of one point (x_2), given its y-coordinate (y_2) and the distance (d) between that point and another given point (x_1, y_1). To do so, the above formula for x_1 should be solved just as a radical equation containing numerical values in place of the other variables. Both sides should be squared; the quantity should be subtracted $(y_2 - y_1)^2$; the square root of both sides should be taken; x_1 should be subtracted to produce:

$$\sqrt{d^2 - (y_2 - y_1)^2} + x_1 = x_2$$

Additional Topics

Volume Formulas

Volume is the capacity of a three-dimensional shape. Volume is useful in determining the space within a certain three-dimensional object. Volume can be calculated for a cube, rectangular prism, cylinder, pyramid, cone, and sphere. By knowing specific dimensions of the objects, the volume of the object is computed with these figures. The units for the volumes of solids can include cubic centimeters, cubic meters, cubic inches, and cubic feet.

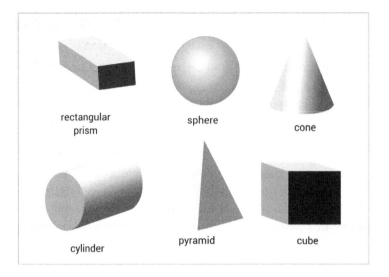

Cube

The cube is the simplest figure for which volume can be determined because all dimensions in a cube are equal. In the following figure, the length, width, and height of the cube are all represented by the variable *a* because these measurements are equal lengths.

The volume of any rectangular, three-dimensional object is found by multiplying its length by its width by its height. In the case of a cube, the length, width, and height are all equal lengths, represented by the variable *a*. Therefore, the equation used to calculate the volume is $(a \times a \times a)$ or a^3. In a real-world example of this situation, if the length of a side of the cube is 3 centimeters, the volume is calculated by utilizing the formula:

$$(3 \times 3 \times 3) = 9 \text{ cm}^3$$

Rectangular Prism

The dimensions of a rectangular prism are not necessarily equal as those of a cube. Therefore, the formula for a rectangular prism recognizes that the dimensions vary and use different variables to represent these lengths. The length, width, and height of a rectangular prism are represented with the variables *a*, *b*, and *c*.

The equation used to calculate volume is length times width times height. Using the variables in the diagram above, this means $a \times b \times c$. In a real-world application of this situation, if *a*=2 cm, *b*=3 cm, and *c*=4 cm, the volume is calculated by utilizing the formula $3 \times 4 \times 5 = 60 \text{ cm}^3$.

Cylinder

Discovering a cylinder's volume requires the measurement of the cylinder's base, length of the radius, and height. The height of the cylinder can be represented with variable *h*, and the radius can be represented with variable *r*.

The formula to find the volume of a cylinder is $\pi r^2 h$. Notice that πr^2 is the formula for the area of a circle. This is because the base of the cylinder is a circle. To calculate the volume of a cylinder, the slices of circles needed to build the entire height of the cylinder are added together. For example, if the radius

is 5 feet and the height of the cylinder is 10 feet, the cylinder's volume is calculated by using the following equation: $\pi 5^2 \times 10$. Substituting 3.14 for π, the volume is 785.4 ft³.

Pyramid

To calculate the volume of a pyramid, the area of the base of the pyramid is multiplied by the pyramid's height by $\frac{1}{3}$. The area of the base of the pyramid is found by multiplying the base length by the base width.

Therefore, the formula to calculate a pyramid's volume is $(L \times W \times H) \div 3$.

Cone

The formula to calculate the volume of a circular cone is similar to the formula for the volume of a pyramid. The primary difference in determining the area of a cone is that a circle serves as the base of a cone. Therefore, the area of a circle is used for the cone's base.

The variable r represents the radius, and the variable h represents the height of the cone. The formula used to calculate the volume of a cone is $\frac{1}{3}\pi r^2 h$. Essentially, the area of the base of the cone is multiplied by the cone's height.

In a real-life example where the radius of a cone is 2 meters and the height of a cone is 5 meters, the volume of the cone is calculated by utilizing the formula:

$$\frac{1}{3}\pi 2^2 \times 5 = 21$$

After substituting 3.14 for π, the volume is 785.4 ft³.

Sphere

The volume of a sphere uses π due to its circular shape.

The length of the radius, r, is the only variable needed to determine the sphere's volume. The formula to calculate the volume of a sphere is $\frac{4}{3}\pi r^3$.

Therefore, if the radius of a sphere is 8 centimeters, the volume of the sphere is calculated by utilizing the formula:

$$\frac{4}{3}\pi(8)^3 = 2{,}143 \; cm^3$$

Right Triangles: Pythagorean Theorem and Trigonometric Ratio

The value of a missing side of a right triangle may be determined two ways. The first way is to apply the Pythagorean Theorem, and the second way is to apply Trigonometric Ratios. The Pythagorean Theorem states that for every right triangle, the square of the length of the hypotenuse is equal to the sum of the squares of the lengths of the remaining two sides. The hypotenuse is the longest side of a right triangle, and is also the side opposite the right angle.

According to the diagram, $a^2 + b^2 = c^2$, where c represents the hypotenuse, and a and b represent the lengths of remaining two sides of the right triangle.

The Pythagorean Theorem may be applied a multitude of ways. For example, a person wishes to build a garden in the shape of a rectangle, having the dimensions of 5 feet by 8 feet. The garden's design includes a diagonal board to separate various types of plants. The Pythagorean Theorem can be used to determine the length of the diagonal board.

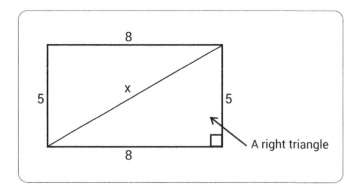

$$a^2 + b^2 = c^2$$

$$5^2 + 8^2 = c^2$$

$$25 + 64 = c^2$$

$$c = \sqrt{89}$$

$$c = 9.43$$

To solve for unknown sides of a right triangle using trigonometric ratios, the sine, cosine, and tangent are required.

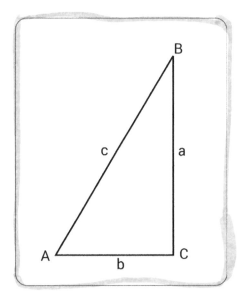

In the image above, angles are denoted by capital letters, and sides are denoted by lowercase letters. When examining angle A, b is the adjacent side, a is the opposite side, and c is the hypotenuse side. The various ratios of the lengths of the sides of the right triangle are used to find the sine, cosine, and tangent of angle A.

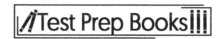
Thus:

$$\sin(A) = \frac{opposite}{hypotenuse}$$

$$\cos(A) = \frac{adjacent}{hypotenuse}$$

$$\tan(A) = \frac{opposite}{adjacent}$$

After substituting variables for the sides of the right triangle:

$$\sin(A) = \frac{a}{c}$$

$$\cos(A) = \frac{b}{c}$$

$$\tan(A) = \frac{a}{b}.$$

As a real-world example, the height of a tree can be discovered by using the information above. Surveying equipment can determine the tree's angle of inclination is 55.3 degrees, and the distance from the tree is 10 feet.

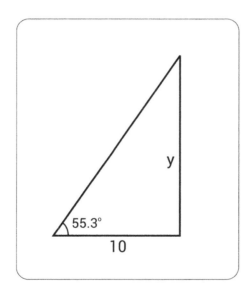

To find the height of the tree, substitute the known values into the trigonometric ratio of tangent:

$$\tan(55.3) = \frac{y}{10}$$

$$10 \times \tan(55.3) = y$$

$$10 \times 1.44418 = y$$

$$y = 14.4418$$

Operations with Complex Numbers

Complex numbers are numbers that have a real component and an imaginary component. An example of a complex number is $3 + 4i$. The real part of this complex number is 3, and the imaginary part of this complex number is $4i$. It is important to note that the imaginary number i is $\sqrt{-1}$. Complex numbers can be added, subtracted, multiplied, and divided.

Adding complex numbers together is similar to adding like terms. If given two complex numbers, students should first add the real components together and then add the imaginary components together. In this way, i is treated like a variable because it is only added or subtracted with other terms that contain i. For example, if asked to simplify $(2 + 4i) + (3 - 5i)$, students should first add the real components together and then add the imaginary components together:

$$(2 + 4i) + (3 - 5i)$$

$$(2 + 3) + (4i + -5i) = 5 - i$$

In addition, if asked to subtract two complex numbers, students should first subtract the real components and then subtract the imaginary components: $(3 + 4i) - (1 + 2i)$ simplifies to:

$$(3 - 1) + (4i - 2i) = 2 + 2i$$

The examples below demonstrate how the imaginary number i is treated when it is raised to a power:

$$i^1 = i$$

$$i^2 = -1$$

$$i^3 = i \times -1 = -i$$

$$i^4 = -1 \times -1 = 1$$

To multiply complex numbers, students should use the FOIL distribution method and combine like terms. FOIL stands for first, outer, inner, and last. For example, when given two expressions of $(2 + 4i)(3 + 2i)$, the student multiplies the first term in each expression (2×3) to get 6. Next, the student multiplies the two outer terms together $(2 \times 2i)$ to get $4i$. The student multiplies the two inner terms together $(4i \times 3)$ to get $12i$. Then the student multiplies the last term of each expression together $(4i \times 2i)$ to get $8i^2$. If using the values described above, $8i^2$ can be further simplified to -8 (since $i^2 = -1$). As a final step, the student combines like terms:

$$6 + 4i + 12i + -8 = -2 + 16i$$

To find the conjugate of a complex number, the sign is changed between the two terms in the denominator. For example, given the complex number $4 + 2i$, the student should change the operation sign in the middle of the two terms from addition to subtraction. Therefore, the complex conjugate becomes $4 - 2i$.

to get a real denominator

To divide complex numbers, the student should multiply by the conjugate of the complex number. The next step is to use the FOIL method in both the numerator and the denominator with the conjugate. For

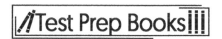

example, when given $\frac{2+2i}{3+i}$, the conjugate of the denominator should be found first. The conjugate of $(3 + i)$ is $(3 - i)$, because the addition sign is changed to a subtraction sign. Given the new expression:

$$\frac{2+2i}{3+i} \times \frac{3-i}{3-i}$$

the student multiplies the two expressions in the numerator using the FOIL distribution method and the two expressions in the denominator using the FOIL distribution method. The numerator simplifies to:

$$(6 - 2i + 6i + -2i^2) = 8 + 4i$$

The denominator simplifies to:

$$(9 - 3i + 3i \pm i^2) = 10$$

As a final step, the student combines like terms:

$$\frac{8 + 4i}{10}$$

which simplifies to:

$$\frac{4 + 2i}{5}$$

Degrees and Radians

Degrees are used to express the size of an angle. A complete circle is represented by 360°, and a half circle is represented by 180°. In addition, a right angle fills one quarter of a circle and is represented by 90°.

Radians are another way to denote angles in terms of π, rather than degrees. A complete circle is represented by 2π radians.

The formula used to convert degrees to radians is:

$$Radians = \frac{degrees \times \pi}{180}$$

For example, to convert 270 degrees to radians:

$$Radians = \frac{270 \times \pi}{180} = 4.71$$

The *arc of a circle* is the distance between two points on the circle. The length of the arc of a circle in terms of *degrees* is easily determined if the value of the central angle is known. The length of the arc is simply the value of the central angle. In this example, the length of the arc of the circle in degrees is 75°.

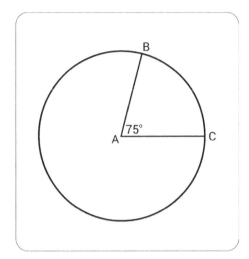

To determine the length of the arc of a circle in *distance*, the student will need to know the values for both the central angle and the radius. This formula is:

$$\frac{central\ angle}{360°} = \frac{arc\ length}{2\pi r}$$

The equation is simplified by cross-multiplying to solve for the arc length.

In the following example, the student should substitute the values of the central angle (75°) and the radius (10 inches) into the equation above to solve for the arc length.

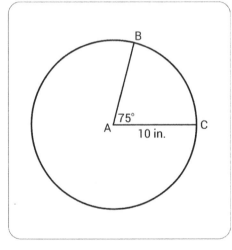

$$\frac{75°}{360°} = \frac{arc\ length}{2(3.14)(10in.)}$$

To solve the equation, first cross-multiply: 4710 = 360(arc length). Next, divide each side of the equation by 360. The result of the formula is that the arc length is 13.1 (rounded). Please note that arc length is often referred to as *s*.

As a special technological note for trigonometric functions, when finding the trigonometric function or an angle on the calculator, make a note using degrees or radians to get the correct value. Whether computing the sine of $\frac{\pi}{6}$ or computing the sine of 30°, the answer should come out to $\frac{1}{2}$. However, there is usually a "Mode" function on the calculator to select either radian or degree.

Circles

The equation used to find the area of a circle is $A = \pi r^2$. For example, if a circle has a radius of 5 centimeters, the area is computed by substituting 5 for the radius: $(5)^2$. Using this reasoning, to find half of the area of a circle, the formula is $A = .5\pi r^2$. Similarly, to find the quarter of an area of a circle, the formula is $A = .25\pi r^2$. To find any fractional area of a circle, a student can use the formula $A = \frac{C}{360}\pi r^2$, where C is the number of degrees of the central angle of the sector. The area of a circle can also be found by using the arc length rather than the degree of the sector. This formula is $A = rs^2$, where s is the arc length and r is the radius of the circle.

A chord is a line that connects two points on a circle's circumference. If the radius and the value of the angle subtended at the center by the chord is known, the formula to find the chord length is:

$$C = 2 \times radius \times \sin\frac{angle}{2}$$

Remember that this formula is based on half the length of the chord, so the radius is doubled to determine the full length of the chord.

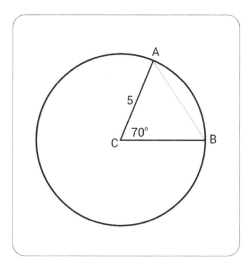

For example, the radius in the diagram above is 5 and the angle is 70 degrees. To find the chord length, plug in the values for the radius and angle to obtain the answer of 5.7.

$$5 \times \sin\frac{70}{2} = 2.87 \times 2 = 5.7$$

Chords that intersect each other at a point within a circle are related. The intersecting chord theorem states that when two chords intersect, each is cut into two portions or segments. The products of the two segments of each respective chord are equal to one another.

Other related concepts for circles include the diameter and circumference. *Circumference* is the distance around a circle. The formula for circumference is $C = 2\pi r$. The *diameter* of a circle is the distance across a circle through its center point. The formula for circumference can also be thought of as $C = dr$ where d is the circle's diameter, since the diameter of a circle is *2r*.

Similarity, Congruence, and Triangles

Triangles are similar if they have the same shape, the same angle measurements, and their sides are proportional to one another. Triangles are congruent if the angles of the triangles are equal in measurement and the sides of the triangles are equal in measurement.

There are five ways to show that a triangle is congruent.

- SSS (Side-Side-Side Postulate): When all three corresponding sides are equal in length, then the two triangles are congruent.

- SAS (Side-Angle-Side Postulate): If a pair of corresponding sides and the angle in between those two sides are equal, then the two triangles are congruent.

- ASA (Angle-Side-Angle Postulate): If a pair of corresponding angles are equal and the side within those angles are equal, then the two triangles are equal.

- AAS (Angle-Angle-Side Postulate): When a pair of corresponding angles for two triangles and a non-included side are equal, then the two triangles are congruent.

- HL (Hypotenuse-Leg Theorem): If two right triangles have the same hypotenuse length, and one of the other sides are also the same length, then the two triangles are congruent.

If two triangles are discovered to be similar or congruent, this information can assist in determining unknown parts of triangles, such as missing angles and sides.

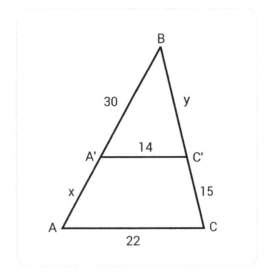

In the triangle shown above, *AC* and *A'C'* are parallel lines. Therefore, *BA* is a transversal that intersects the two parallel lines. The corresponding angles *BA'C'* and *BAC* are congruent. In a similar way, *BC* is also a transversal. Therefore, angle *BC'A'* and *BCA* are congruent. If two triangles have two congruent angles, the triangles are similar. If the triangles are similar, their corresponding sides are proportional.

Therefore, the following equation is established:

$$\frac{30 + x}{30} = \frac{22}{14} = \frac{y + 15}{y}$$

$$\frac{30 + x}{30} = \frac{22}{14}$$

$$x = 17.1$$

$$\frac{22}{14} = \frac{y + 15}{y}$$

$$y = 26.25$$

The example below involves the question of congruent triangles. The first step is to examine whether the triangles are congruent. If the triangles are congruent, then the measure of a missing angle can be found.

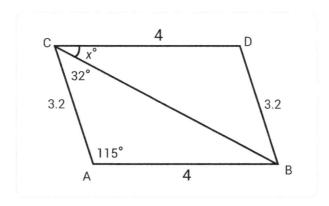

The above diagram provides values for angle measurements and side lengths in triangles *CAB* and *CDB*. Note that side *CA* is 3.2 and side *DB* is 3.2. Side *CD* is 4 and side *AB* is 4. Furthermore, line *CB* is congruent to itself by the reflexive property. Therefore, the two triangles are congruent by SSS (Side-Side-Side). Because the two triangles are congruent, all of the corresponding parts of the triangles are also congruent. Therefore, angle *x* is congruent to the inside of the angle for which a measurement is not provided in Triangle *CAB*. Thus, $115° + 32° = 147°$. A triangle measures 180°, therefore $180° - 147° = 33°$. *Angle x* = 33°, because the two triangles are reversed.

Complementary Angle Theorem

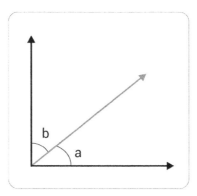

Two angles are complementary if the sum of the two angles equals 90°.

In the above diagram $Angle\ a + Angle\ b = 90°$. Therefore, the two angles are complementary. Certain trigonometric rules are also associated with complementary angles.

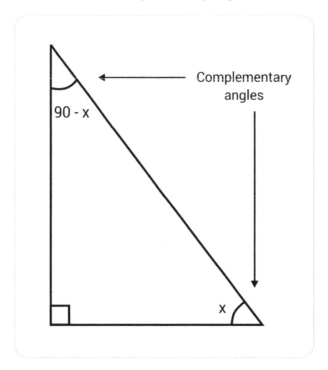

In the diagram above of a right triangle, if Angle *A* and Angle *C* are determined to be complementary angles, then certain relationships can be stated between the trigonometry of those angles.

$$sin(90° - x) = cos\ x$$

$$cos(90° - x) = sin\ x$$

For example, the sine of 80 degrees equals the cosine of $(90° - 80°)$, which is the cos $(10°)$.

This is true because the sine of an angle in a right triangle is equal to the cosine of its complement. Sine is known as the conjunction of cosine, and cosine is known as the conjunction of sine.

Examples:

1. $cos5° = sin\ x°$?
2. $sin(90° - x) = ?\ sin(90° - x) =?$

For problem number 1, the student should remember that $sin(90° - x) = cos\ x$. Cos 5° would be the same as $sin(90 - 5)°$. Therefore, $cos5° = sin85°$.

For problem number 2, the student would use the same fact that $sin(90° - x)° = cos\ x$.

An *acute angle* is an angle that is less than 90°. If Angle *A* and Angle *B* are acute angles of a right triangle, then $sinA = cosB$. Therefore, the sine of any acute angle in a right triangle is equal to the cosine of its complement, and the cosine of any acute angle is equal to the sine of its complement.

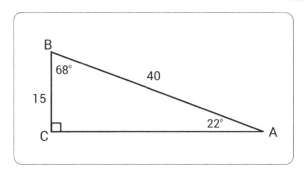

The example above is a right triangle. If only the value of angle *BAC* (which is 22°) was provided, the student would be able to figure out the value for angle *CBA* (68°) by knowing that a triangle is made up of 180°(180° − 90° − 22° = 68°). From the information given about acute angles on the previous page, the following statement is true:

Sine (angle *BAC*) = $\frac{15}{40}$, which is equivalent to the Cos (angle *CBA*) = $\frac{15}{40}$

Circles on the Coordinate Plane

If a circle is placed on the coordinate plane with the center of the circle at the origin (0,0), then point (*x, y*) is a point on the circle. Furthermore, the line extending from the center to point (*x, y*) is the radius, or *r*. By applying the Pythagorean Theorem ($a^2 + b^2 = c^2$), it can be stated that $x^2 + y^2 = r^2$. However, the center of the circle does not always need to be on the origin of the coordinate plane.

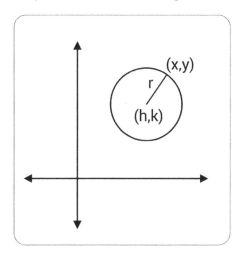

In the diagram above, the center of the circle is noted by (*h, k*). By applying the distance formula, the equation becomes:

$$= \sqrt{(x - h)^2 + (y - k)^2}$$

When squaring both sides of the equation, the result is the standard form of a circle with the center (h, k) and radius r. Namely, $r^2 = (x - h)^2 + (y - k)^2$, where r = radius and center = (h, k). The following examples may be solved by using this information:

Example: Graph the equation $-x^2 + y^2 = 25$

To graph this equation, first note that the center of the circle is (0, 0). The radius is the positive square root of 25 or 5.

Example: Find the equation for the circle below.

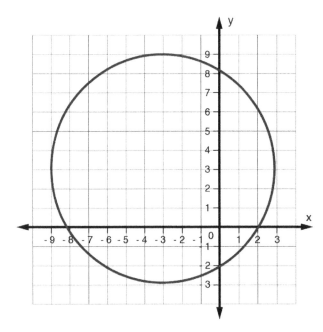

To find the equation for the circle, note that its center is not zero. Therefore, to find the circle's center, draw vertical and horizontal diameters to examine where they intersect. The center is located at point: (-3, 3). Next, count the number of spaces from the center to the outside of the circle. This number is 6. Therefore, 6 is the radius. Finally, plug in the numbers that are known into the standard equation for a circle:

$$36 = \left(x - (-3)\right)^2 + (y - 3)^2$$

or

$$36 = (x + 3)^2 + (y - 3)^2$$

It is possible to determine whether a point lies on a circle or not within the coordinate plane. For example, a circle has a center of (2, -5), and a radius of 6 centimeters. The first step is to apply the equation of a circle, which is $r^2 = (x - h)^2 + (y - k)^2$, where r = radius and the center = (h, k). Next, substitute the numbers for the center point and the number for the radius. This action simplifies the equation to:

$$36 = (x - 2)^2 + (y + 5)^2$$

Note that the radius of 6 was squared to get 36.

To prove that the point (2, -1) lies on the circle, apply the equation of the circle that was just used and input the values of (2, -1) for *x* and *y* in the equation.

$$36 = (x - 2)^2 + (y + 5)^2$$

$$36 = (2 - 2)^2 + (1 + 5)^2$$

$$36 = (0)^2 + (6)^2$$

$$36 = 36$$

Because the left side of the equation equals the right side of the equation, point (2, 1) lies on the given circle.

Practice Questions

1. Store brand coffee beans cost $1.23 per pound. A local coffee bean roaster charges $1.98 per 1 ½ pounds. How much more would 5 pounds from the local roaster cost than 5 pounds of the store brand?
    a. $0.55
    b. $1.55
    c. $1.45
    d. $0.45

2. Which of the following formulas would correctly calculate the perimeter of a legal-sized piece of paper that is 14 inches long and $8\frac{1}{2}$ inches wide?
    a. $P = 14 + 8\frac{1}{2}$
    b. $P = 14 + 8\frac{1}{2} + 14 + 8\frac{1}{2}$
    c. $P = 14 \times 8\frac{1}{2}$
    d. $P = 14 \times \frac{17}{2}$

3. What are the coordinates of the focus of the parabola $y = -9x^2$?
    a. $(-3, 0)$
    b. $\left(-\frac{1}{36}, 0\right)$
    c. $(0, -3)$
    d. $\left(0, -\frac{1}{36}\right)$

4. Where does the point (–3, –4) lie in relation to the circle with the equation $(x)^2 + (y)^2 = 25$?
    a. Inside of the circle.
    b. Outside of the circle.
    c. On the circle.
    d. There is not enough information to tell.

5.. The total perimeter of a rectangle is 36 cm. If the length is 12 cm, what is the width?

Answer Explanations

1. D: $0.45

List the givens.

$$Store\ coffee = \$1.23/lbs$$

$$Local\ roaster\ coffee = \$1.98/1.5\ lbs$$

Calculate the cost for 5 lbs of store brand.

$$\frac{\$1.23}{1\ lbs} \times 5\ lbs = \$6.15$$

Calculate the cost for 5 lbs of the local roaster.

$$\frac{\$1.98}{1.5\ lbs} \times 5\ lbs = \$6.60$$

Subtract to find the difference in price for 5 lbs.

$$\begin{array}{r} \$6.60 \\ - \$6.15 \\ \hline \$0.45 \end{array}$$

2. B: The perimeter of a rectangle is the sum of all four sides. Therefore, the answer is:

$$P = 14 + 8\frac{1}{2} + 14 + 8\frac{1}{2}$$

$$14 + 14 + 8 + \frac{1}{2} + 8 + \frac{1}{2} = 45 \text{ square inches}$$

3. D: A parabola of the form $y = \frac{1}{4f}x^2$ has a focus $(0, f)$. Because $y = -9x^2$, set $-9 = \frac{1}{4f}$. Solving this equation for f results in $f = -\frac{1}{36}$. Therefore, the coordinates of the focus are $\left(0, -\frac{1}{36}\right)$.

4. C: Plug in the values for x and y to discover that the solution works, which is:

$$(-3)^2 + (-4)^2 = 25$$

Choices *A* and *B* are not the correct answers since the solution works. Choice *D* is not the correct answer because there is enough information to tell where the given point lies on the circle.

5.

The formula for the perimeter of a rectangle is P=2l+2w, where P is the perimeter, l is the length, and w is the width. The first step is to substitute all of the data into the formula:

$$36 = 2(12) + 2W$$

Simplify by multiplying 2x12:

$$36 = 24 + 2W$$

Simplifying this further by subtracting 24 on each side, which gives:

$$36 - 24 = 24 - 24 + 2W$$

$$12 = 2W$$

Divide by 2:

$$6 = W$$

The width is 6 cm. Remember to test this answer by substituting this value into the original formula:

$$36 = 2(12) + 2(6)$$

Reading

U.S. Literature Passage #1

Objective: Read and comprehend an excerpt from U.S. Literature.

Questions 1-6 are based on the following passage:

The following passage is an excerpt from *The Curious Case of Benjamin Button*, F.S. Fitzgerald, 1922

As long ago as 1860 it was the proper thing to be born at home. At present, so I am told, the high gods of medicine have decreed that the first cries of the young shall be uttered upon the anesthetic air of a hospital, preferably a fashionable one. So young Mr. and Mrs. Roger Button were fifty years ahead of style when they decided, one day in the summer of 1860, that their first baby should be born in a hospital. Whether this anachronism had any bearing upon the astonishing history I am about to set down will never be known.

I shall tell you what occurred, and let you judge for yourself.

The Roger Buttons held an enviable position, both social and financial, in ante-bellum Baltimore. They were related to the This Family and the That Family, which, as every Southerner knew, entitled them to membership in that enormous peerage which largely populated the Confederacy. This was their first experience with the charming old custom of having babies— Mr. Button was naturally nervous. He hoped it would be a boy so that he could be sent to Yale College in Connecticut, at which institution Mr. Button himself had been known for four years by the somewhat obvious nickname of "Cuff."

On the September morning <u>consecrated</u> to the enormous event he arose nervously at six o'clock dressed himself, adjusted an impeccable stock, and hurried forth through the streets of Baltimore to the hospital, to determine whether the darkness of the night had borne in new life upon its bosom.

When he was approximately a hundred yards from the Maryland Private Hospital for Ladies and Gentlemen he saw Doctor Keene, the family physician, descending the front steps, rubbing his hands together with a washing movement—as all doctors are required to do by the unwritten ethics of their profession.

Mr. Roger Button, the president of Roger Button & Co., Wholesale Hardware, began to run toward Doctor Keene with much less dignity than was expected from a Southern gentleman of that picturesque period. "Doctor Keene!" he called. "Oh, Doctor Keene!"

The doctor heard him, faced around, and stood waiting, a curious expression settling on his harsh, medicinal face as Mr. Button drew near.

"What happened?" demanded Mr. Button, as he came up in a gasping rush. "What was it? How is she? A boy? Who is it? What—"

"Talk sense!" said Doctor Keene sharply. He appeared somewhat irritated.

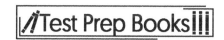

"Is the child born?" begged Mr. Button.

Doctor Keene frowned. "Why, yes, I suppose so—after a fashion." Again he threw a curious glance at Mr. Button.

1. What major event is about to happen in this story?
   a. Mr. Button is about to go to a funeral.
   b. Mr. Button's wife is about to have a baby.
   c. Mr. Button is getting ready to go to the doctor's office.
   d. Mr. Button is about to go shopping for new clothes.

2. What kind of tone does the above passage have?
   a. Nervous and Excited
   b. Sad and Angry
   c. Shameful and Confused
   d. Grateful and Joyous

3. What is the meaning of the word "consecrated" in paragraph 4?
   a. Numbed
   b. Chained
   c. Dedicated
   d. Moved

4. What does the author mean to do by adding the following statement?

   "rubbing his hands together with a washing movement—as all doctors are required to do by the unwritten ethics of their profession."

   a. Suggesting that Mr. Button is tired of the doctor.
   b. Trying to explain the detail of the doctor's profession.
   c. Hinting to readers that the doctor is an unethical man.
   d. Giving readers a visual picture of what the doctor is doing.

5. Which of the following best describes the development of this passage?
   a. It starts in the middle of a narrative in order to transition smoothly to a conclusion.
   b. It is a chronological narrative from beginning to end.
   c. The sequence of events is backwards—we go from future events to past events.
   d. To introduce the setting of the story and its characters.

6. Which of the following is an example of an imperative sentence?
   a. "Oh, Doctor Keene!"
   b. "Talk sense!"
   c. "Is the child born?"
   d. "Why, yes, I suppose so—"

Literature Passage #2

Objective: Read and comprehend an excerpt from U.S. Literature.

Questions 7-12 are based on the following passage:

The following is an excerpt from "The Story of An Hour," Kate Chopin, 1894

Knowing that Mrs. Mallard was afflicted with heart trouble, great care was taken to break to her as gently as possible the news of her husband's death.

It was her sister Josephine who told her, in broken sentences; veiled hints that revealed in half concealing. Her husband's friend Richards was there, too, near her. It was he who had been in the newspaper office when intelligence of the railroad disaster was received, with Brently Mallard's name leading the list of "killed." He had only taken the time to assure himself of its truth by a second telegram, and had hastened to forestall any less careful, less tender friend in bearing the sad message.

She did not hear the story as many women have heard the same, with a paralyzed inability to accept its significance. She wept at once, with sudden, wild abandonment, in her sister's arms. When the storm of grief had spent itself she went away to her room alone. She would have no one follow her.

There stood, facing the open window, a comfortable, roomy armchair. Into this she sank, pressed down by a physical exhaustion that haunted her body and seemed to reach into her soul.

She could see in the open square before her house the tops of trees that were all aquiver with the new spring life. The delicious breath of rain was in the air. In the street below a peddler was crying his wares. The notes of a distant song which some one was singing reached her faintly, and countless sparrows were twittering in the eaves.

There were patches of blue sky showing here and there through the clouds that had met and piled one above the other in the west facing her window.

She sat with her head thrown back upon the cushion of the chair, quite motionless, except when a sob came up into her throat and shook her, as a child who has cried itself to sleep continues to sob in its dreams.

She was young, with a fair, calm face, whose lines bespoke repression and even a certain strength. But now here was a dull stare in her eyes, whose gaze was fixed away off yonder on one of those patches of blue sky. It was not a glance of reflection, but rather indicated a suspension of intelligent thought.

There was something coming to her and she was waiting for it, fearfully. What was it? She did not know; it was too subtle and elusive to name. But she felt it, creeping out of the sky, reaching toward her through the sounds, the scents, and color that filled the air.

Now her bosom rose and fell tumultuously. She was beginning to recognize this thing that was approaching to possess her, and she was striving to beat it back with her will—as powerless as her two white slender hands would have been. When she abandoned herself a little whispered

word escaped her slightly parted lips. She said it over and over under her breath: "free, free, free!" The vacant stare and the look of terror that had followed it went from her eyes. They stayed keen and bright. Her pulses beat fast, and the coursing blood warmed and relaxed every inch of her body.

She did not stop to ask if it were or were not a monstrous joy that held her. A clear and exalted perception enabled her to dismiss the suggestion as trivial. She knew that she would weep again when she saw the kind, tender hands folded in death; the face that had never looked save with love upon her, fixed and gray and dead. But she saw beyond that bitter moment a long procession of years to come that would belong to her absolutely. And she opened and spread her arms out to them in welcome.

7. What point of view is the above passage told in?
   a. First person
   b. Second person
   c. Third person omniscient
   d. Third person limited

8. What kind of irony are we presented with in this story?
   a. The way Mrs. Mallard reacted to her husband's death.
   b. The way in which Mr. Mallard died.
   c. The way in which the news of her husband's death was presented to Mrs. Mallard.
   d. The way in which nature is compared with death in the story.

9. What is the meaning of the word "elusive" in paragraph 9?
   a. Horrible
   b. Indefinable
   c. Quiet
   d. Joyful

10. What is the best summary of the passage above?
   a. Mr. Mallard, a soldier during World War I, is killed by the enemy and leaves his wife widowed.
   b. Mrs. Mallard understands the value of friendship when her friends show up for her after her husband's death.
   c. Mrs. Mallard combats mental illness daily and will perhaps be sent to a mental institution soon.
   d. Mrs. Mallard, a newly widowed woman, finds unexpected relief in her husband's death.

11. What is the tone of this story?
   a. Confused
   b. Joyful
   c. Depressive
   d. All of the above

12. What is the meaning of the word "tumultuously" in paragraph 10?
   a. Orderly
   b. Unashamedly
   c. Violently
   d. Calmly

History Passage #1

Objective: Read and comprehend an excerpt written from a historical perspective.

Questions 13-18 are based upon the following passage:

"MANKIND being originally equals in the order of creation, the equality could only be destroyed by some subsequent circumstance; the distinctions of rich, and poor, may in a great measure be accounted for, and that without having recourse to the harsh ill sounding names of oppression and avarice. Oppression is often the consequence, but seldom or never the means of riches; and though avarice will preserve a man from being necessitously poor, it generally makes him too timorous to be wealthy.

But there is another and greater distinction for which no truly natural or religious reason can be assigned, and that is, the distinction of men into KINGS and SUBJECTS. Male and female are the distinctions of nature, good and bad the distinctions of heaven; but how a race of men came into the world so exalted above the rest, and distinguished like some new species, is worth enquiring into, and whether they are the means of happiness or of misery to mankind.

In the early ages of the world, according to the scripture chronology, there were no kings; the consequence of which was there were no wars; it is the pride of kings which throw mankind into confusion Holland without a king hath enjoyed more peace for this last century than any of the monarchical governments in Europe. Antiquity favors the same remark; for the quiet and rural lives of the first patriarchs hath a happy something in them, which vanishes away when we come to the history of Jewish royalty.

Government by kings was first introduced into the world by the Heathens, from whom the children of Israel copied the custom. It was the most prosperous invention the Devil ever set on foot for the promotion of idolatry. The Heathens paid divine honors to their deceased kings, and the Christian world hath improved on the plan by doing the same to their living ones. How impious is the title of sacred majesty applied to a worm, who in the midst of his splendor is crumbling into dust!

As the exalting one man so greatly above the rest cannot be justified on the equal rights of nature, so neither can it be defended on the authority of scripture; for the will of the Almighty, as declared by Gideon and the prophet Samuel, expressly disapproves of government by kings. All anti-monarchical parts of scripture have been very smoothly glossed over in monarchical governments, but they undoubtedly merit the attention of countries, which have their governments yet to form. "Render unto Caesar the things which are Caesar's" is the scripture doctrine of courts, yet it is no support of monarchical government, for the Jews at that time were without a king, and in a state of vassalage to the Romans.

Near three thousand years passed away from the Mosaic account of the creation, till the Jews under a national delusion requested a king. Till then their form of government (except in extraordinary cases, where the Almighty interposed) was a kind of republic administered by a judge and the elders of the tribes. Kings they had none, and it was held sinful to acknowledge any being under that title but the Lord of Hosts. And when a man seriously reflects on the idolatrous homage which is paid to the persons of Kings,

he need not wonder, that the Almighty ever jealous of his honor, should disapprove of a form of government which so impiously invades the prerogative of heaven.

Excerpt From: Thomas Paine. "Common Sense."

13. According to passage, what role does avarice, or greed, play in poverty?
   a. It can make a man very wealthy.
   b. It is the consequence of wealth.
   c. Avarice can prevent a man from being poor, but too fearful to be very wealthy.
   d. Avarice is what drives a person to be very wealthy

14. Of these distinctions, which does the author believe to be beyond natural or religious reason?
   a. Good and bad
   b. Male and female
   c. Human and animal
   d. King and subjects

15. According to the passage, what are the Heathens responsible for?
   a. Government by kings
   b. Quiet and rural lives of patriarchs
   c. Paying divine honors to their living kings
   d. Equal rights of nature

16. Which of the following best states Paine's rationale for the denouncement of monarchy?
   a. It is against the laws of nature.
   b. It is against the equal rights of nature and is denounced in scripture.
   c. Despite scripture, a monarchal government is unlawful.
   d. Neither the law nor scripture denounce monarchy.

17. Based on the passage, what is the best definition of the word *idolatrous*?
   a. Worshipping heroes
   b. Being deceitful
   c. Sinfulness
   d. Engaging in illegal activities

18. What is the essential meaning of lines 41-44?
   And when a man seriously reflects on the idolatrous homage which is paid to the persons of Kings, he need not wonder, that the Almighty ever jealous of his honor, should disapprove of a form of government which so impiously invades the prerogative of heaven.

   a. God would disapprove of the irreverence of a monarchical government.
   b. With careful reflection, men should realize that heaven is not promised.
   c. God will punish those that follow a monarchical government.
   d. Belief in a monarchical government cannot coexist with belief in God.

History Passage #2

Objective: Read and comprehend an excerpt written from a historical perspective.

The following excerpt is from the article "The Lancashire Witches 1612–2012," by Robert Poole. Please read it and answer questions 19–24.

Four hundred years ago, in 1612, the north-west of England was the scene of England's biggest peacetime witch trial: the trial of the Lancashire witches. Twenty people, mostly from the Pendle area of Lancashire, were imprisoned in the castle as witches. Ten were hanged, one died in gaol, one was sentenced to stand in the pillory, and eight were acquitted. The 2012 anniversary sees a small flood of commemorative events, including works of fiction by Blake Morrison, Carol Ann Duffy, and Jeanette Winterson. How did this witch trial come about, and what accounts for its enduring fame?

We know so much about the Lancashire Witches because the trial was recorded in unique detail by the clerk of the court, Thomas Potts, who published his account soon afterwards as *The Wonderful Discovery of Witches in the County of Lancaster*. I have recently published a modern-English edition of this book, together with an essay piecing together what we know of the events of 1612. It has been a fascinating exercise, revealing how Potts carefully edited the evidence, and also how the case against the "witches" was constructed and manipulated to bring about a spectacular show trial. It all began in mid-March when a pedlar from Halifax named John Law had a frightening encounter with a poor young woman, Alizon Device, in a field near Colne. He refused her request for pins and there was a brief argument during which he was seized by a fit that left him with "his head … drawn awry, his eyes and face deformed, his speech not well to be understood; his thighs and legs stark lame." We can now recognize this as a stroke, perhaps triggered by the stressful encounter. Alizon Device was sent for and surprised all by confessing to the bewitching of John Law and then begged for forgiveness.

When Alizon Device was unable to cure the pedlar, the local magistrate, Roger Nowell was called in. Characterized by Thomas Potts as "God's justice" he was alert to instances of witchcraft, which were regarded by the Lancashire's puritan-inclined authorities as part of the cultural rubble of "popery"—Roman Catholicism—long overdue to be swept away at the end of the county's very slow protestant reformation. "With weeping tears" Alizon explained that she had been led astray by her grandmother, "old Demdike," well-known in the district for her knowledge of old Catholic prayers, charms, cures, magic, and curses. Nowell quickly interviewed Alizon's grandmother and mother, as well as Demdike's supposed rival, "old Chattox" and her daughter Anne. Their panicky attempts to explain themselves and shift the blame to others eventually only ended up incriminating them, and the four were sent to Lancaster gaol in early April to await trial at the summer assizes. The initial picture revealed was of a couple of poor, marginal local families in the forest of Pendle with a longstanding reputation for magical powers, which they had occasionally used at the request of their wealthier neighbours. There had been disputes but none of these were part of ordinary village life. Not until 1612 did any of this come to the attention of the authorities.

The net was widened still further at the end of April when Alizon's younger brother James and younger sister Jennet, only nine years old, came up between them with a story about a "great meeting of witches" at their grandmother's house, known as Malkin Tower. This meeting was presumably to discuss the plight of those arrested and the threat of further arrests, but

according to the evidence extracted from the children by the magistrates, a plot was hatched to blow up Lancaster castle with gunpowder, kill the gaoler, and rescue the imprisoned witches. It was, in short, a conspiracy against royal authority to rival the gunpowder plot of 1605—something to be expected in a county known for its particularly strong underground Roman Catholic presence.

Those present at the meeting were mostly family members and neighbours, but they also included Alice Nutter, described by Potts as "a rich woman [who] had a great estate, and children of good hope: in the common opinion of the world, of good temper, free from envy or malice." Her part in the affair remains mysterious, but she seems to have had Catholic family connections, and may have been one herself, providing an added motive for her to be prosecuted.

This article (The Lancaster Witches 1612-2012) was originally published in *The Public Domain Review* under a Creative Commons Attribution-ShareAlike 3.0. If you wish to reuse it, please see: http://publicdomainreview.org/legal/

19. What's the point of this passage, and why did the author write it?
    a. The author is documenting a historic witchcraft trial while uncovering/investigating the role of suspicion and anti-Catholicism in the events.
    b. The author seeks long-overdue reparations for the ancestors of those accused and executed for witchcraft in Lancashire.
    c. The author is educating the reader about actual occult practices of the 1600s.
    d. The author argues that the Lancashire witch trials were more brutal than the infamous Salem trials.

20. Which term best captures the meaning of the author's use of "enduring" in the first paragraph?
    a. Un-original
    b. Popular
    c. Wicked
    d. Circumstantial

21. What textual information is present within the passage that most lends itself to the author's credibility?
    a. His prose is consistent with the time.
    b. This is a reflective passage; the author doesn't need to establish credibility.
    c. The author cites specific quotes.
    d. The author has published a modern account of the case and has written on the subject before.

22. What might the following excerpt suggest about the trial or, at the very least, Thomas Potts' account of the trial(s)?
    "It has been a fascinating exercise, revealing how Potts carefully edited the evidence, and also how the case against the 'witches' was constructed and manipulated to bring about a spectacular show trial."

    a. The events were so grand that the public was allowed access to such a spectacular set of cases.
    b. Sections may have been exaggerated or stretched to create notoriety on an extraordinary case.
    c. Evidence was faked, making the trial a total farce.
    d. The trial was corrupt from the beginning.

23. Which statement best describes the political atmosphere of the 1600s that influenced the Alizon Device witch trial/case?
    a. Fear of witches was prevalent during this period.
    b. Magistrates were seeking ways to cement their power during this period of unrest.
    c. In a highly superstitious culture, the Protestant church and government were highly motivated to root out any potential sources that could undermine the current regime.
    d. Lancashire was originally a prominent area for pagan celebration, making the modern Protestants very weary of whispers of witchcraft and open to witch trials to resolve any potential threats to Christianity.

24. Which best describes the strongest "evidence" used in the case against Alizon and the witches?
    a. Knowledge of the occult and witchcraft
    b. "Spectral evidence"
    c. Popular rumors of witchcraft and Catholic association
    d. Self-incriminating speech

History Passage #3

Objective: Read and comprehend an excerpt written from a historical perspective.

The following passage is from "The Rise of Natural History Museums," by Oliver Cummings Farrington. Please read it and answer questions 25–30.

A desire to preserve objects of nature which aroused special interest or possessed unusual powers may be presumed to have been an instinct of the earliest man. We may imagine the cave man storing in his cave the bright gem, or curious seed, or rare animal skin which attracted his attention and, perchance, urging upon his descendants the desirability of preserving it. Such instincts are undoubtedly possessed by barbarous tribes. But such hoards have no permanent value or maintenance as long as there is a lack of a fixed habitation or of a social organization sufficiently strong to pass them from one generation to another. Hence, it may be noted in passing, an essential condition for the existence of museums is a sufficiently civilized and permanent state of society to preserve objects from generation to generation.

In the life of the ancient Egyptians conditions making toward the preservation of natural objects doubtless became more favorable than had previously been the case, since there are preserved to us from their time many objects of their art which were originally objects of nature. While material which they prized now occupies an honored place in our museums and their civilization was instrumental in preserving it to us, there is no evidence, so far as I know, that they undertook the collection and preservation of natural objects for their own sake.

The Greeks gave us the word museum, but that they ever established a museum in the modern sense seems very unlikely. Whatever their practice may have been regarding the preservation and exhibition of works of art, it seems quite certain that they carried on little, if any, effort of this kind with regard to nature. Alexander the Great, about 325 BC, is said to have gathered together many animals and plants in order that they might be studied by Aristotle, "the father of natural history," but so far as we know no effort was made to preserve these specimens to later times. The first record of placing natural history specimens on exhibition is said to be made when Hanno, a Carthaginian, somewhat before Alexander's time, procured skins of gorillas in Africa and put them in the temple of Astarte. We also know that the monstrous horns of wild

bulls which had occasioned great devastation in Macedonia were hung in the temple of Hercules by order of King Philip.

The Romans seem, like the Greeks, not to have taken much interest in the preservation of natural objects, at least as far as any record has reached the present time. We know that emperors and other individuals possessed collections of statues and other works of art, and among these we find occasional mention of the preservation of so-called "natural curiosities," such as bones of giants or peculiar human skeletons, but that any broad interest in nature existed which led to efforts to preserve and study its forms we have no record. Stray sources of information tell us of a crocodile, found in attempting to discover the sources of the Nile, being preserved in the temple of Isis at Cesarea, also that a large piece of the root of the cinnamon tree was kept in a golden vessel in one of the temples at Rome. Pliny relates that the bones of a sea monster, probably a whale, "to which Andromeda was exposed," were preserved at Joppa and afterwards brought to Rome. Suetonious says that the Emperor Augustus had a collection of natural curiosities in his palace.

One reason suggested by Beckmann for the rarity of collections of natural objects among ancient peoples was the lack of knowledge of satisfactory means of preserving such as were perishable. The preservative virtues of what was then called "spirit of wine," but which we now know as alcohol, seem to have been but little known, and only immersion in salt brine or a covering with wax or honey served at that time for the preservation of perishable materials.

The great institute of Alexandria in Egypt, founded in the third century BC, is generally spoken of as being the first natural-history museum of antiquity, but while this had botanical and zoological gardens, there is little reason to suppose that it was a museum of nature in the modern sense. The name museum in that institution was applied to a portion set apart for the study of sciences, and indicated rather a place of study than one for exhibition of objects.

25. Which statement best captures the goal of this excerpt?
    a. The author argues that the ancient Greeks, not the Egyptians, invented museums.
    b. The author is reflecting on the history of museums while detailing his own visits to museums.
    c. The author seeks to sketch the history behind modern museums by discussing ancient roots.
    d. The author wants to prove that actual museums didn't evolve until the 1850s.

26. Which best encapsulates the core reasoning behind the author's idea, "Hence, it may be noted in passing, an essential condition for the existence of museums is a sufficiently civilized and permanent state of society to preserve objects from generation to generation"?
    a. A refined level of sophistication is needed to understand and pass down the knowledge of artifacts within a museum.
    b. Stable civil conditions and a learned society are key for museum preservation and education; if the city is unstable, museum interest and artifacts are threatened.
    c. Museums can't exist as a nomadic, or traveling practice; the artifacts won't be preserved.
    d. The continuation of museums relies on educating future generations.

27. What seems to be the main criteria for a site to be considered a museum in the modern sense?
    a. A location that contain artifacts that are archaic in nature and significant to specific events
    b. A central location that houses artifacts
    c. The wealth to purchase items and display them safely without risk of damage or theft
    d. A central location that displays artifacts for the general public to learn and enjoy

28. Which of the following could best replace the term *virtues* in the fifth paragraph?
    a. Properties
    b. Corrections
    c. Understanding
    d. Abilities

29. What can be inferred from the following detail the author offers in the text?
    "Alexander the Great, about 325 BC, is said to have gathered together many animals and plants in order that they might be studied by Aristotle, 'the father of natural history,' but so far as we know no effort was made to preserve these specimens to later time."

    a. Aristotle was the best expert on natural history of his time.
    b. Collections like this led to the discovery of many exotic animals.
    c. Natural history and sciences were actively being studied in ancient times.
    d. Alexander the Great was seeking to build his own museum.

30. The author has decided to include a section that details how the "natural curiosities" that were often attributed to myths and legends were actually misidentified dinosaur bones, or other natural objects. Where would this information fit best?
    a. In a new paragraph, after the second paragraph
    b. In the fourth paragraph, before the last sentence
    c. Just above the last paragraph
    d. In the fourth paragraph, after the second sentence

Science Passage #1

Objective: Read and comprehend an excerpt written from a scientific perspective.

Questions 31-36 are based on the following passage:

In the quest to understand existence, modern philosophers must question if humans can fully comprehend the world. Classical western approaches to philosophy tend to hold that one can understand something, be it an event or object, by standing outside of the phenomena and observing it. It is then by unbiased observation that one can grasp the details of the world. This seems to hold true for many things. Scientists conduct experiments and record their findings, and thus many natural phenomena become comprehendible. However, several of these observations were possible because humans used tools in order to make these discoveries.

This may seem like an extraneous matter. After all, people invented things like microscopes and telescopes in order to enhance their capacity to view cells or the movement of stars. While humans are still capable of seeing things, the question remains if human beings have the capacity to fully observe and see the world in order to understand it. It would not be an impossible stretch to argue that what humans see through a microscope is not the exact thing itself, but a human interpretation of it.

This would seem to be the case in the "Business of the Holes" experiment conducted by Richard Feynman. To study the way electrons behave, Feynman set up a barrier with two holes and a plate. The plate was there to indicate how many times the electrons would pass through the hole(s). Rather than casually observe the electrons acting under

normal circumstances, Feynman discovered that electrons behave in two totally different ways depending on whether or not they are observed. The electrons that were observed had passed through either one of the holes or were caught on the plate as particles. However, electrons that weren't observed acted as waves instead of particles and passed through both holes. This indicated that electrons have a dual nature. Electrons seen by the human eye act like particles, while unseen electrons act like waves of energy.

This dual nature of the electrons presents a conundrum. While humans now have a better understanding of electrons, the fact remains that people cannot entirely perceive how electrons behave without the use of instruments. We can only observe one of the mentioned behaviors, which only provides a partial understanding of the entire function of electrons. Therefore, we're forced to ask ourselves whether the world we observe is objective or if it is subjectively perceived by humans. Or, an alternative question: can man understand the world only through machines that will allow them to observe natural phenomena?

Both questions humble man's capacity to grasp the world. However, those ideas don't consider that many phenomena have been proven by human beings without the use of machines, such as the discovery of gravity. Like all philosophical questions, whether man's reason and observation alone can understand the universe can be approached from many angles.

31. The word *extraneous* in paragraph two can be best interpreted as referring to which one of the following?
    a. Indispensable
    b. Bewildering
    c. Superfluous
    d. Exuberant

32. What is the author's motivation for writing the passage?
    a. To bring to light an alternative view on human perception by examining the role of technology in human understanding.
    b. To educate the reader on the latest astroparticle physics discovery and offer terms that may be unfamiliar to the reader.
    c. To argue that humans are totally blind to the realities of the world by presenting an experiment that proves that electrons are not what they seem on the surface.
    d. To reflect on opposing views of human understanding.

33. Which of the following most closely resembles the way in which paragraph four is structured?
    a. It offers one solution, questions the solution, and then ends with an alternative solution.
    b. It presents an inquiry, explains the details of that inquiry, and then offers a solution.
    c. It presents a problem, explains the details of that problem, and then ends with more inquiry.
    d. It gives a definition, offers an explanation, and then ends with an inquiry.

34. For the classical approach to understanding to hold true, which of the following must be required?
    a. A telescope
    b. A recording device
    c. Multiple witnesses present
    d. The person observing must be unbiased

35. Which best describes how the electrons in the experiment behaved like waves?
    a. The electrons moved up and down like actual waves.
    b. The electrons passed through both holes and then onto the plate.
    c. The electrons converted to photons upon touching the plate.
    d. Electrons were seen passing through one hole or the other.

36. The author mentions "gravity" in the last paragraph in order to do what?
    a. To show that different natural phenomena test man's ability to grasp the world.
    b. To prove that since man has not measured it with the use of tools or machines, humans cannot know the true nature of gravity.
    c. To demonstrate an example of natural phenomena humans discovered and understand without the use of tools or machines.
    d. To show an alternative solution to the nature of electrons that humans have not thought of yet.

Science Passage #2

Objective: Read and comprehend an excerpt written from a scientific perspective.

The following passage is an excerpt from The Myth of the Birth of the Hero, A Psychological Interpretation of Mythology, *by Otto Rank. Read it and answer questions 37–42.*

> The prominent civilized nations—the Babylonians and Egyptians, the Hebrews and Hindus, the Persians, the Greeks and the Romans, as well as the Teutons and others—all began at an early stage to glorify their national heroes—mythical princes and kings, founders of religions, dynasties, empires, or cities—in a number of poetic tales and legends. The history of the birth and of the early life of these personalities came to be especially invested with fantastic features, which in different nations—even though widely separated by space and entirely independent of each other—present a baffling similarity or, in part, a literal correspondence. Many investigators have long been impressed with this fact, and one of the chief problems of mythological research still consists in the elucidation of the reason for the extensive analogies in the fundamental outlines of mythical tales, which are rendered still more puzzling by the unanimity in certain details and their reappearance in most of the mythical groupings.
>
> The mythological theories, aiming at the explanation of these remarkable phenomena, are, in a general way, as follows:
>
> 1. The "Idea of the People," propounded by Adolf Bastian. This theory assumes the existence of elemental ideas, so that the unanimity of the myths is a necessary sequence of the uniform disposition of the human mind and the manner of its manifestation, which within certain limits is identical at all times and in all places. This interpretation was urgently advocated by Adolf Bauer as accounting for the wide distribution of the hero myths.

2. The explanation by original community, first applied by Theodor Benfey to the widely distributed parallel forms of folklore and fairy tales. Originating in a favorable locality (India), these tales were first accepted by the primarily related (Indo-Germanic) peoples, then continued to grow while retaining the common primary traits, and ultimately radiated over the entire earth. This mode of explanation was first adapted to the wide distribution of the hero myths by Rudolf Schubert.

3. The modern theory of migration, or borrowing, according to which individual myths originate from definite peoples (especially the Babylonians) and are accepted by other peoples through oral tradition (commerce and traffic) or through literary influences. The modern theory of migration and borrowing can be readily shown to be merely a modification of Benfey's theory, necessitated by newly discovered and irreconcilable material. This profound and extensive research of modern investigations has shown that India, rather than Babylonia, may be regarded as the first home of the myths. Moreover, the tales presumably did not radiate from a single point, but traveled over and across the entire inhabited globe. This brings into prominence the idea of the interdependence of mythological structures, an idea which was generalized by Braun as the basic law of the nature of the human mind: Nothing new is ever discovered as long as it is possible to copy. The theory of elemental ideas, so strenuously advocated by Bauer over a quarter of a century ago, is unconditionally declined by the most recent investigators (Winckler, Stucken), who maintain the migration theory.

There is really no such sharp contrast between the various theories or their advocates, for the concept of elemental ideas does not interfere with the claims of primary common possession or of migration. Furthermore, the ultimate problem is not whence and how the material reached a certain people; the question is: Where did it come from to begin with? All these theories would explain only the variability and distribution of the myths, but not their origin. Even Schubert, the most inveterate opponent of Bauer's view, acknowledges this truth, by stating that all these manifold sagas date back to a single very ancient prototype. But he is unable to tell us anything of the origin of this prototype. Bauer likewise inclines to this mediating view; he points out repeatedly that in spite of the multiple origin of independent tales, it is necessary to concede a most extensive and ramified borrowing, as well as an original community of the concepts in related peoples.

37. Which term best defines *elucidation* as it's used in the first paragraph of the passage?
    a. Definition
    b. Specification
    c. Ramification
    d. Explanation

38. Based on the title of his work and the context of the selected passage, which statement may serve as the best explanation for why Rank is studying mythology?
    a. Psychologist Carl Jung proved the concept of archetypes which seems to suggest universal concepts and ideas.
    b. The proliferation of common mythic structures around the world and cultures suggest shared, fundamental human ideas and values. To study these myths is to study the core of human thought.
    c. The study of the mythic hero may uncover the true origins of the first leaders.
    d. Studying mythology enables Rank to analyze how conflicts can be mitigated across various cultures, helping him develop new psychological analysis strategies and therapies.

39. Which statement provides an example that would correlate the following theory from the main passage?

> "The 'Idea of the People,' propounded by Adolf Bastian. This theory assumes the existence of elemental ideas, so that the unanimity of the myths is a necessary sequence of the uniform disposition of the human mind and the manner of its manifestation, which within certain limits is identical at all times and in all places."

a. Human beings have the need to understand their origins, hence the presence of creation myths.
b. The idea of winter as a cold season needed explanation; therefore, myths developed to analyze and interpret the natural phenomena, creating stories that account for seasonal change.
c. Heroes exemplify power and strength.
d. The Germanic and Nordic people wanted to understand why chaos exists in the world; the giants, beings of chaos and destruction, explained how natural phenomena occurred.

40. Reading through the body of the text, why is it appropriate that the mentioned theories are in fact called *theories*?

a. They are scientific explanations.
b. There is no reason; this was an artistic choice.
c. While very insightful, these theories as of yet cannot be officially proven; however, they are likely.
d. These are all competing ideas.

41. What are some insights that can be drawn from the following description of the third theory of mythic origins statement that can also relate to the rest of the passage?

> "This brings into prominence the idea of the interdependence of mythological structures, an idea which was generalized by Braun as the basic law of the nature of the human mind: Nothing new is ever discovered as long as it is possible to copy."

a. Humans lack originality; there are no new ideas.
b. The cultural minds of human beings evolved from a set of basic ideas.
c. Human culture is interdependent on one another; there is no unique culture but a sundered mythic cycle that once was universal.
d. Recurring mythic structures seem to be reiterations of shared human experiences/stories, used over and over but modified throughout various cultures.

42. Which answer best exemplifies the mentioned theory/explanation in the following description of the second theory of mythic origins?

> "The explanation by original community, first applied by Theodor Benfey to the widely distributed parallel forms of folklore and fairy tales. Originating in a favorable locality (India), these tales were first accepted by the primarily related (Indo-Germanic) peoples, then continued to grow while retaining the common primary traits, and ultimately radiated over the entire earth."

a. A tribe begins trading with another tribe. Through this interaction, different myths are shared, including the story of the hero.
b. A tribe has a legend of how the sky was formed. At some point, the tribe splits into different tribes, and each retains the myth. Some of the newer tribes attribute sky creation to different gods or tricksters.
c. The myth of Aeneas and Dido explains the ancient rivalry between Carthage and Rome.
d. There is a myth of a world flood in many cultures.

Science Passage #3

Objective: Read and comprehend an excerpt written from a scientific perspective.

The following excerpt is from "The Culture Emergence of Man", by Alton Howard Thompson. Read it and answer questions 43–47.

Dr. Frank Baker, of the National Zoological Park, at Washington, kindly wrote that "One monkey in our collection, when annoyed by visitors, will throw anything, from a feed-pan to a handful of sawdust, at an offender. One Cebus tried to pick cockroaches out of the cracks in the floor with a straw, when too small for his fingers; but beyond this there has been nothing observed that could be considered as the using of an object as a tool or weapon." Other correspondents said the same thing—there were no actions on the part of the quadrumana, that they had observed, that could be taken as indicative of intelligent action. Prof. R. L. Garner, in his most interesting book on the "Speech of Apes and Monkeys," who spent some time in the wild country of the Gaboon on purpose to pursue his studies, observes that "animals may be taught to do many things in a mechanical way and without any motives that relate to the actions." His pet chimpanzee tried to drive nails, use the saw, etc., but could not manage it, nor even the use of the club to crush his sugar-cane. Of the gorilla he says: "As to his throwing sticks or stones at enemies, there is nothing to verify it and much to contradict it. It is a mere freak of fancy. Neither the chimpanzee nor the gorilla close the hand to strike nor use any weapon but the hands and teeth."

From this evidence at first hand, we must conclude that the use of extraneous substances by animals, especially the quadrumana, is purely automatic and imitative, and not to be considered as rational action at all. Whatever they have learned has been by reason of contact with man and the result of imitation and training. We must conclude, further, that the use of tools and weapons, even preceding the intelligent conformation of them for definite purposes, marked the differentiation of primitive man from the animal branch, and accompanied indeed was the cause of the psychic emergence. The moment that the primitive man-ape employed extraneous substances intelligently, with a purpose, he ceased to be an ape and became a man. We must believe that the use of tools and the psychic emergence were coincident and interdependent.

As M. de Pressense says, in his "Study of Origins" (352): "The first tool fashioned by man asserted his royalty over nature. Thus, the tool is man's true scepter; whether made of flint or wood or anything else—it is the result of thought. This is why the animal, guided by instinct, can affect marvels of construction by the use of its own limbs, but it never makes a tool. A monkey may have chanced one day to lean upon a stick, but he did not cut nor shape the stick nor hand it down to posterity, that they might improve upon it."

The struggles of the first ape-man to maintain life amid the hostile surroundings in which he found himself are fraught with peculiar interest, and are indeed almost pathetic when we consider the great odds that were against him in the fight. His natural weapons of defense, the jaws and teeth, were being reduced with a rapidity that must soon have brought about his extinction but for the correlative development of the grasping powers of the hand, which enabled him to employ and supplement his natural organs with the extra natural resources around him. And then it followed, as this grasping power enabled him to use a club or a stone, that some superior individual made a conscious effort to employ these weapons with more precision and initiate new purposes, and that he thereby learned to think. This was the divine spark that awakened mental life, which acted as a stimulus on the motor nerve centers, and these centers were enlarged by the effort to think. Then, as the brain grew, he could think more, and as he thought more his brain grew, and he became a man. If you will pardon the solecism, this primeval man might have said with Descartes (much to Descartes's surprise, probably, by the application), "Cogito; ergo sum"—I think; therefore I exist. It is an old and true saying that "man is the wisest of animals because of his hands"—a pre-Darwinian appreciation of correlated development that was prophetic.

We must begin, then, with the first efforts of primitive man to isolate himself from the animal world by the use of his hands, and the exercise of that manual power which distinguishes him from the rest of the animal kingdom and made him its master. We must consider, however, that primeval man was at first incapable of manufacturing implements and weapons from the materials around him, and was only capable of using in a simple way the gifts of nature as they came from her hands without any modification whatever. Kindly nature furnished him with these resources to supplement the waning powers of his natural organs, which were being rapidly modified in the process of his evolution. Primitive man utilized the simple things that nature furnished ready to his hand and they were sufficient for his needs. The primeval life of the human race must therefore be considered first in the light of what nature provided for practical use and which was of vital importance in the struggle for existence. But these were sufficient to give him the balance of power, and he lived. To this primeval man nature was kind and beneficent, and nursed and nurtured him to the full development of the maturity of the race, in his civilized descendants. From a mere animal she enabled him to develop into the godlike being who dominates the earth.

43. Which example of observation and/or discovery disproves the following excerpt from the passage? "From this evidence at first hand, we must conclude that the use of extraneous substances by animals, especially the quadrumana, is purely automatic and imitative, and not to be considered as rational action at all."

   a. Gorillas, chimpanzees, and orangutans have successfully learned sign language from humans.
   b. Modern chimpanzees have been discovered using stone tools in the wild.
   c. A Capuchin monkey learned to use a screwdriver for several advanced tasks after watching a video.
   d. Captive apes have taught their offspring sign language without being instructed by human trainers.

44. The term *quadrumana* is used multiple times in the text. What is the most accurate definition for this word that can be inferred from the text?
   a. Primates with four grasping hands
   b. Primates that evolved in the last forty thousand years
   c. Primates with five digits on each hand
   d. The classification of monkeys and great apes that spend time both in the trees and on the ground

45. What's the significance of the "psychic emergence" mentioned in the second paragraph?
   a. It represents the development of the frontal lobe, crucial for processing higher thoughts.
   b. It represents the development of mental powers needed to shape weapons.
   c. It represents the cognizance that items can be shaped and used for a purpose.
   d. It represents the development of free will.

46. Which answer best explains how Descartes' famous quote, "Cogito ergo sum," which Thompson translates to "I think; therefore, I exist," illustrates the author's main point?
    a. Being able to think is crucial to tool use.
    b. Self-awareness enables individuals to see nature objectively, so they will intentionally use resources and make decisions to advance their own existence.
    c. Self-awareness is what led to humans developing higher mental functions and tool making.
    d. The ability to understand natural resources and use was the beginning of the concept of individual thought and ambition to create.

47. Which best describes how the term *scepter* serves as an appropriate metaphor in Thompson's quote from M. de Pressense's *Study of Origins*, "Thus the tool is man's true scepter"?
    a. It symbolizes kingship and the ability to create and settle the world.
    b. It is a precious tool.
    c. It illustrates the inherited right for humans to rule the world due to their intelligence.
    d. It serves to illustrate man's independence and power to control natural elements (tools).

Writing

Questions 1–9 are based on the following passage:

While all dogs (1) <u>descend through gray wolves</u>, it's easy to notice that dog breeds come in a variety of shapes and sizes. With such a (2) <u>drastic range of traits, appearances and body types</u> dogs are one of the most variable and adaptable species on the planet. (3) <u>But why so many differences.</u> The answer is that humans have actually played a major role in altering the biology of dogs. (4) <u>This was done through a process called selective breeding.</u>

(5) <u>Selective breeding which is also called artificial selection is the processes</u> in which animals with desired traits are bred in order to produce offspring that share the same traits. In natural selection, (6) <u>animals must adapt to their environments increase their chance of survival.</u> Over time, certain traits develop in animals that enable them to thrive in these environments. Those animals with more of these traits, or better versions of these traits, gain an (7) <u>advantage over others of their species.</u> Therefore, the animal's chances to mate are increased and these useful (8) <u>genes are passed into their offspring.</u> With dog breeding, humans select traits that are desired and encourage more of these desired traits in other dogs by breeding dogs that already have them.

The reason for different breeds of dogs is that there were specific needs that humans wanted to fill with their animals. For example, scent hounds are known for their extraordinary ability to track game through scent. These breeds are also known for their endurance in seeking deer and other prey. Therefore, early hunters took dogs that displayed these abilities and bred them to encourage these traits. Later, these generations took on characteristics that aided these desired traits. (9) <u>For example, Bloodhounds</u> have broad snouts and droopy ears that fall to the ground when they smell. These physical qualities not only define the look of the Bloodhound, but also contribute to their amazing tracking ability. The broad snout is able to define and hold onto scents longer than many other breeds. The long floppy hears serve to collect and hold the scents the earth holds so that the smells are clearer and able to be distinguished.

1. Which of the following would be the best choice for this sentence (reproduced below)?

While all dogs (1) <u>descend through gray wolves</u>, it's easy to notice that dog breeds come in a variety of shapes and sizes.

a. NO CHANGE
b. descend by gray wolves
c. descend from gray wolves
d. descended through gray wolves

2. Which of the following would be the best choice for this sentence (reproduced below)?

With such a (2) <u>drastic range of traits, appearances and body types</u>, dogs are one of the most variable and adaptable species on the planet.

a. NO CHANGE
b. drastic range of traits, appearances, and body types,
c. drastic range of traits and appearances and body types,
d. drastic range of traits, appearances, as well as body types,

3. Which of the following would be the best choice for this sentence (reproduced below)?

(3) <u>But why so many differences.</u>

a. NO CHANGE
b. But are there so many differences?
c. But why so many differences are there.
d. But why so many differences?

4. Which of the following would be the best choice for this sentence (reproduced below)?

(4) <u>This was done through a process called selective breeding.</u>

a. NO CHANGE
b. This was done, through a process called selective breeding.
c. This was done, through a process, called selective breeding.
d. This was done through selective breeding, a process.

5. Which of the following would be the best choice for this sentence (reproduced below)?

(5) <u>Selective breeding which is also called artificial selection is the processes</u> in which animals with desired traits are bred in order to produce offspring that share the same traits.

a. NO CHANGE
b. Selective breeding, which is also called artificial selection is the processes
c. Selective breeding which is also called, artificial selection, is the processes
d. Selective breeding, which is also called artificial selection, is the processes

6. Which of the following would be the best choice for this sentence (reproduced below)?

In natural selection, (6) <u>animals must adapt to their environments increase their chance of survival.</u>

   a. NO CHANGE
   b. animals must adapt to their environments to increase their chance of survival.
   c. animals must adapt to their environments, increase their chance of survival.
   d. animals must adapt to their environments, increasing their chance of survival.

7. Which of the following would be the best choice for this sentence (reproduced below)?

Those animals with more of these traits, or better versions of these traits, gain an (7) <u>advantage over others of their species.</u>

   a. NO CHANGE
   b. advantage over others, of their species.
   c. advantages over others of their species.
   d. advantage over others.

8. Which of the following would be the best choice for this sentence (reproduced below)?

Therefore, the animal's chances to mate are increased and these useful (8) <u>genes are passed into their offspring.</u>

   a. NO CHANGE
   b. genes are passed onto their offspring.
   c. genes are passed on to their offspring.
   d. genes are passed within their offspring.

9. Which of the following would be the best choice for this sentence (reproduced below)?

(9) <u>For example, Bloodhounds</u> have broad snouts and droopy ears that fall to the ground when they smell.

   a. NO CHANGE
   b. For example, Bloodhounds,
   c. For example Bloodhounds
   d. For example, bloodhounds

Questions 10–18 are based on the following passage:

I'm not alone when I say that it's hard to pay attention sometimes. I can't count how many times I've sat in a classroom, lecture, speech, or workshop and (10) <u>been bored to tears or rather sleep</u>. (11) <u>Usually I turn to doodling in order to keep awake</u>. This never really helps; I'm not much of an artist. Therefore, after giving up on drawing a masterpiece, I would just concentrate on keeping my eyes open and trying to be attentive. This didn't always work because I wasn't engaged in what was going on.

(12) <u>Sometimes in particularly dull seminars,</u> I'd imagine comical things going on in the room or with the people trapped in the room with me. Why? (13) <u>Because I wasn't invested in what was</u>

going on I wasn't motivated to listen. I'm not going to write about how I conquered the difficult task of actually paying attention in a difficult or unappealing class—it can be done, sure. I have sat through the very epitome of boredom (in my view at least) several times and come away learning something. (14) <u>Everyone probably has had to at one time do this.</u> What I want to talk about is that profound moment when curiosity is sparked (15) <u>in another person drawing them to pay attention to what is before them</u> and expand their knowledge.

What really makes people pay attention? (16) <u>Easy it's interest.</u> This doesn't necessarily mean (17) <u>embellishing subject matter drawing people's attention.</u> This won't always work. However, an individual can present material in a way that is clear to understand and actually engages the audience. Asking questions to the audience or class will make them a part of the topic at hand. Discussions that make people think about the content and (18) <u>how it applies to there lives world and future is key.</u> If math is being discussed, an instructor can explain the purpose behind the equations or perhaps use real-world applications to show how relevant the topic is. When discussing history, a lecturer can prompt students to imagine themselves in the place of key figures and ask how they might respond. The bottom line is to explore the ideas rather than just lecture. Give people the chance to explore material from multiple angles, and they'll be hungry to keep paying attention for more information.

10. Which of the following would be the best choice for this sentence (reproduced below)?

    I can't count how many times I've sat in a classroom, lecture, speech, or workshop and (10) <u>been bored to tears or rather sleep.</u>

    a. NO CHANGE
    b. been bored to, tears, or rather sleep.
    c. been bored, to tears or rather sleep.
    d. been bored to tears or, rather, sleep.

11. Which of the following would be the best choice for this sentence (reproduced below)?

    (11) <u>Usually I turn to doodling in order to keep awake.</u>

    a. NO CHANGE
    b. Usually, I turn to doodling in order to keep awake.
    c. Usually I turn to doodling, in order, to keep awake.
    d. Usually I turned to doodling in order to keep awake.

12. Which of the following would be the best choice for this sentence (reproduced below)?

    (12) <u>Sometimes in particularly dull seminars,</u> I'd imagine comical things going on in the room or with the people trapped in the room with me.

    a. NO CHANGE
    b. Sometimes, in particularly, dull seminars,
    c. Sometimes in particularly dull seminars
    d. Sometimes in particularly, dull seminars,

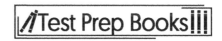

13. Which of the following would be the best choice for this sentence (reproduced below)?

(13) Because I wasn't invested in what was going on I wasn't motivated to listen.

a. NO CHANGE
b. Because I wasn't invested, in what was going on, I wasn't motivated to listen.
c. Because I wasn't invested in what was going on. I wasn't motivated to listen.
d. I wasn't motivated to listen because I wasn't invested in what was going on.

14. Which of the following would be the best choice for this sentence (reproduced below)?

(14) Everyone probably has had to at one time do this.

a. NO CHANGE
b. Everyone probably has had to, at one time. Do this.
c. Everyone's probably had to do this at some time.
d. At one time everyone probably has had to do this.

15. Which of the following would be the best choice for this sentence (reproduced below)?

What I want to talk about is that profound moment when curiosity is sparked (15) in another person drawing them to pay attention to what is before them and expand their knowledge.

a. NO CHANGE
b. in another person, drawing them to pay attention
c. in another person; drawing them to pay attention to what is before them.
d. in another person, drawing them to pay attention to what is before them.

16. Which of the following would be the best choice for this sentence (reproduced below)?

(16) Easy it's interest.

a. NO CHANGE
b. Easy it is interest.
c. Easy. It's interest.
d. Easy—it's interest.

17. Which of the following would be the best choice for this sentence (reproduced below)?

This doesn't necessarily mean (17) embellishing subject matter drawing people's attention.

a. NO CHANGE
b. embellishing subject matter which draws people's attention.
c. embellishing subject matter to draw people's attention.
d. embellishing subject matter for the purpose of drawing people's attention.

18. Which of the following would be the best choice for this sentence (reproduced below)?

Discussions that make people think about the content and (18) how it applies to there lives world and future is key.

a. NO CHANGE
b. how it applies to their lives, world, and future is key.
c. how it applied to there lives world and future is key.
d. how it applies to their lives, world and future is key.

Questions 19–27 are based on the following passage:

Since the first discovery of dinosaur bones, (19) scientists has made strides in technological development and methodologies used to investigate these extinct animals. We know more about dinosaurs than ever before and are still learning fascinating new things about how they looked and lived. However, one has to ask, (20) how if earlier perceptions of dinosaurs continue to influence people's understanding of these creatures? Can these perceptions inhibit progress towards further understanding of dinosaurs?

(21) The biggest problem with studying dinosaurs is simply that there are no living dinosaurs to observe. All discoveries associated with these animals are based on physical remains. To gauge behavioral characteristics, scientists cross-examine these (22) finds with living animals that seem similar in order to gain understanding. While this method is effective, these are still deductions. Some ideas about dinosaurs can't be tested and confirmed simply because humans can't replicate a living dinosaur. For example, a Spinosaurus has a large sail, or a finlike structure that grows from its back. Paleontologists know this sail exists and have ideas for the function of (23) the sail however they are uncertain of which idea is the true function. Some scientists believe (24) the sail serves to regulate the Spinosaurus' body temperature and yet others believe its used to attract mates. Still, other scientists think the sail is used to intimidate other predatory dinosaurs for self-defense. These are all viable explanations, but they are also influenced by what scientists know about modern animals. (25) Yet, it's quite possible that the sail could hold a completely unique function.

While it's (26) plausible, even likely that dinosaurs share many traits with modern animals, there is the danger of overattributing these qualities to a unique, extinct species. For much of the early nineteenth century, when people first started studying dinosaur bones, the assumption was that they were simply giant lizards. (27) For the longest time this image was the prevailing view on dinosaurs, until evidence indicated that they were more likely warm blooded. Scientists have also discovered that many dinosaurs had feathers and actually share many traits with modern birds.

19. Which of the following would be the best choice for this sentence (reproduced below)?

Since the first discovery of dinosaur bones, (19) <u>scientists has made strides in technological development and methodologies used to investigate</u> these extinct animals.

a. NO CHANGE
b. scientists has made strides in technological development, and methodologies, used to investigate
c. scientists have made strides in technological development and methodologies used to investigate
d. scientists, have made strides in technological development and methodologies used, to investigate

20. Which of the following would be the best choice for this sentence (reproduced below)?

However, one has to ask, (20) <u>how if earlier perceptions of dinosaurs</u> continue to influence people's understanding of these creatures?

a. NO CHANGE
b. how perceptions of dinosaurs
c. how, if, earlier perceptions of dinosaurs
d. whether earlier perceptions of dinosaurs

21. Which of the following would be the best choice for this sentence (reproduced below)?

(21) <u>The biggest problem with studying dinosaurs is simply that there are no living dinosaurs to observe.</u>

a. NO CHANGE
b. The biggest problem with studying dinosaurs is simple, that there are no living dinosaurs to observe.
c. The biggest problem with studying dinosaurs is simple. There are no living dinosaurs to observe.
d. The biggest problem with studying dinosaurs, is simply that there are no living dinosaurs to observe.

22. Which of the following would be the best choice for this sentence (reproduced below)?

To gauge behavioral characteristics, scientists cross-examine these (22) <u>finds with living animals that seem similar in order to gain understanding.</u>

a. NO CHANGE
b. finds with living animals to explore potential similarities.
c. finds with living animals to gain understanding of similarities.
d. finds with living animals that seem similar, in order, to gain understanding.

23. Which of the following would be the best choice for this sentence (reproduced below)?

Paleontologists know this sail exists and have ideas for the function of (23) <u>the sail however they are</u> <u>uncertain of which idea is the true function.</u>

a. NO CHANGE
b. the sail however, they are uncertain of which idea is the true function.
c. the sail however they are, uncertain, of which idea is the true function.
d. the sail; however, they are uncertain of which idea is the true function.

24. Which of the following would be the best choice for this sentence (reproduced below)?

Some scientists believe (24) <u>the sail serves to regulate the Spinosaurus' body temperature and yet</u> <u>others believe its used to attract mates.</u>

a. NO CHANGE
b. the sail serves to regulate the Spinosaurus' body temperature, yet others believe it's used to attract mates.
c. the sail serves to regulate the Spinosaurus' body temperature and yet others believe it's used to attract mates.
d. the sail serves to regulate the Spinosaurus' body temperature however others believe it's used to attract mates.

25. Which of the following would be the best choice for this sentence (reproduced below)?

(25) <u>Yet, it's quite possible</u> that the sail could hold a completely unique function.

a. NO CHANGE
b. Yet, it's quite possible,
c. It's quite possible,
d. Its quite possible

26. Which of the following would be the best choice for this sentence (reproduced below)?

While it's (26) <u>plausible, even likely that dinosaurs share many</u> traits with modern animals, there is the danger of over attributing these qualities to a unique, extinct species.

a. NO CHANGE
b. plausible, even likely that, dinosaurs share many
c. plausible, even likely, that dinosaurs share many
d. plausible even likely that dinosaurs share many

27. Which of the following would be the best choice for this sentence (reproduced below)?

(27) <u>For the longest time this image was the prevailing view on dinosaurs</u>, until evidence indicated that they were more likely warm blooded.

a. NO CHANGE
b. For the longest time this was the prevailing view on dinosaurs
c. For the longest time, this image, was the prevailing view on dinosaurs
d. For the longest time this was the prevailing image of dinosaurs

Questions 28–36 are based on the following passage:

Everyone has heard the (28) <u>idea of the end justifying the means; that would be Weston's philosophy</u>. Weston is willing to cross any line, commit any act no matter how heinous, to achieve success in his goal. (29) <u>Ransom is reviled by this fact, seeing total evil in Weston's plan.</u> To do an evil act in order (30) <u>to gain a result that's supposedly good would ultimately warp the final act.</u> (31) <u>This opposing viewpoints immediately distinguishes Ransom as the hero.</u> In the conflict with Un-man, Ransom remains true to his moral principles, someone who refuses to be compromised by power. Instead, Ransom makes it clear that by allowing such processes as murder and lying dictate how one attains a positive outcome, (32) <u>the righteous goal becomes corrupted.</u> The good end would not be truly good, but a twisted end that conceals corrupt deeds.

(33) <u>This idea of allowing necessary evils to happen, is very tempting, it is what Weston fell prey to.</u> (34) <u>The temptation of the evil spirit Un-man ultimately takes over Weston and he is possessed.</u> However, Ransom does not give into temptation. He remains faithful to the truth of what is right and incorrect. This leads him to directly face Un-man for the fate of Perelandra and its inhabitants.

Just as Weston was corrupted by the Un-man, (35) <u>Un-man after this seeks to tempt the Queen of Perelandra to darkness.</u> Ransom must literally (36) <u>show her the right path, to accomplish this, he does this based on the same principle as the "means to an end" argument</u>—that good follows good, and evil follows evil. Later in the plot, Weston/Un-man seeks to use deceptive reasoning to turn the queen to sin, pushing the queen to essentially ignore Melildil's rule to satisfy her own curiosity. In this sense, Un-man takes on the role of a false prophet, a tempter. Ransom must shed light on the truth, but this is difficult; his adversary is very clever and uses brilliant language. Ransom's lack of refinement heightens the weight of Un-man's corrupted logic, and so the Queen herself is intrigued by his logic.

Based on an excerpt from *Perelandra* by C.S. Lewis

28. Which of the following would be the best choice for this sentence (reproduced below)?

Everyone has heard the (28) idea of the end justifying the means; that would be Weston's philosophy.

a. NO CHANGE
b. idea of the end justifying the means; this is Weston's philosophy.
c. idea of the end justifying the means, this is the philosophy of Weston
d. idea of the end justifying the means. That would be Weston's philosophy.

29. Which of the following would be the best choice for this sentence (reproduced below)?

(29) Ransom is reviled by this fact, seeing total evil in Weston's plan.

a. NO CHANGE
b. Ransom is reviled by this fact; seeing total evil in Weston's plan.
c. Ransom, is reviled by this fact, seeing total evil in Weston's plan.
d. Ransom reviled by this, sees total evil in Weston's plan.

30. Which of the following would be the best choice for this sentence (reproduced below)?

To do an evil act in order (30) to gain a result that's supposedly good would ultimately warp the final act.

a. NO CHANGE
b. for an outcome that's for a greater good would ultimately warp the final act.
c. to gain a final act would warp its goodness.
d. to achieve a positive outcome would ultimately warp the goodness of the final act.

31. Which of the following would be the best choice for this sentence (reproduced below)?

(31) This opposing viewpoints immediately distinguishes Ransom as the hero.

a. NO CHANGE
b. This opposing viewpoints immediately distinguishes Ransom, as the hero.
c. This opposing viewpoint immediately distinguishes Ransom as the hero.
d. Those opposing viewpoints immediately distinguishes Ransom as the hero.

32. Which of the following would be the best choice for this sentence (reproduced below)?

Instead, Ransom makes it clear that by allowing such processes as murder and lying dictate how one attains a positive outcome, (32) the righteous goal becomes corrupted.

a. NO CHANGE
b. the goal becomes corrupted and no longer righteous.
c. the righteous goal becomes, corrupted.
d. the goal becomes corrupted, when once it was righteous.

33. Which of the following would be the best choice for this sentence (reproduced below)?

(33) This idea of allowing necessary evils to happen, is very tempting, it is what Weston fell prey to.

a. NO CHANGE
b. This idea of allowing necessary evils to happen, is very tempting. This is what Weston fell prey to.
c. This idea, allowing necessary evils to happen, is very tempting, it is what Weston fell prey to.
d. This tempting idea of allowing necessary evils to happen is what Weston fell prey to.

34. Which of the following would be the best choice for this sentence (reproduced below)?

(34) The temptation of the evil spirit Un-man ultimately takes over Weston and he is possessed.

a. NO CHANGE
b. The temptation of the evil spirit Un-man ultimately takes over and possesses Weston.
c. Weston is possessed as a result of the temptation of the evil spirit Un-man ultimately, who takes over.
d. The temptation of the evil spirit Un-man takes over Weston and he is possessed ultimately.

35. Which of the following would be the best choice for this sentence (reproduced below)?

Just as Weston was corrupted by the Un-man, (35) Un-man after this seeks to tempt the Queen of Perelandra to darkness.

a. NO CHANGE
b. Un-man, after this, would tempt the Queen of Perelandra
c. Un-man, after this, seeks to tempt the Queen of Perelandra
d. Un-man then seeks to tempt the Queen of Perelandra

36. Which of the following would be the best choice for this sentence (reproduced below)?

Ransom must literally (36) show her the right path, to accomplish this, he does this based on the same principle as the "means to an end" argument—that good follows good, and evil follows evil.

a. NO CHANGE
b. show her the right path. To accomplish this, he uses the same principle as the "means to an end" argument
c. show her the right path; to accomplish this he uses the same principle as the "means to an end" argument
d. show her the right path, to accomplish this, the same principle as the "means to an end" argument is applied

Questions 37–44 are based on the following passage:

(37) What's clear about the news is today is that the broader the media the more ways there are to tell a story. Even if different news groups cover the same story, individual newsrooms can interpret or depict the story differently than other counterparts. Stories can also change depending on the type of (38) media in question incorporating different styles and unique ways to approach the news. (39) It is because of these respective media types that ethical and news-

related subject matter can sometimes seem different or altered. But how does this affect the narrative of the new story?

I began by investing a written newspaper article from the Baltimore Sun. Instantly striking are the bolded Headlines. (40) These are clearly meant for direct the viewer to the most exciting and important stories the paper has to offer. What was particularly noteworthy about this edition was that the first page dealt with two major ethical issues. (41) On a national level there was a story on the evolving Petraeus scandal involving his supposed affair. The other article was focused locally in Baltimore, a piece questioning the city's Ethic's Board and their current director. Just as a television newscaster communicates the story through camera and dialogue, the printed article applies intentional and targeted written narrative style. More so than any of the mediums, news article seems to be focused specifically on a given story without need to jump to another. Finer details are usually expanded on (42) in written articles, usually people who read newspapers or go online for web articles want more than a quick blurb. The diction of the story is also more precise and can be either straightforward or suggestive (43) depending in earnest on the goal of the writer. However, there's still plenty of room for opinions to be inserted into the text.

Usually, all news (44) outlets have some sort of bias, it's just a question of how much bias clouds the reporting. As long as this bias doesn't withhold information from the reader, it can be considered credible. However, an over use of bias, opinion, and suggestive language can rob readers of the chance to interpret the news events for themselves.

37. Which of the following would be the best choice for this sentence (reproduced below)?

(37) What's clear about the news today is that the broader the media the more ways there are to tell a story.

a. NO CHANGE
b. What's clear, about the news today, is that the broader the media
c. What's clear about today's news is that the broader the media
d. The news today is broader than earlier media

38. Which of the following would be the best choice for this sentence (reproduced below)?

Stories can also change depending on the type of (38) media in question incorporating different styles and unique ways to approach the news.

a. NO CHANGE
b. media in question; each incorporates unique styles and unique
c. media in question. To incorporate different styles and unique
d. media in question, incorporating different styles and unique

39. Which of the following would be the best choice for this sentence (reproduced below)?

(39) It is because of these respective media types that ethical and news-related subject matter can sometimes seem different or altered.

a. NO CHANGE
b. It is because of these respective media types, that ethical and news-related subject matter, can sometimes seem different or altered.
c. It is because of these respective media types, that ethical and news-related subject matter can sometimes seem different or altered.
d. It is because of these respective media types that ethical and news-related subject matter can sometimes seem different. Or altered.

40. Which of the following would be the best choice for this sentence (reproduced below)?

(40) These are clearly meant for direct the viewer to the most exciting and important stories the paper has to offer.

a. NO CHANGE
b. These are clearly meant for the purpose of giving direction to the viewer
c. These are clearly meant to direct the viewer
d. These are clearly meant for the viewer to be directed

41. Which of the following would be the best choice for this sentence (reproduced below)?

(41) On a national level there was a story on the evolving Petraeus scandal involving his supposed affair.

a. NO CHANGE
b. On a national level a story was there
c. On a national level; there was a story
d. On a national level, there was a story

42. Which of the following would be the best choice for this sentence (reproduced below)?

Finer details are usually expanded on (42) in written articles, usually people who read newspapers or go online for web articles want more than a quick blurb.

a. NO CHANGE
b. in written articles. People who usually
c. in written articles, usually, people who
d. in written articles usually people who

43. Which of the following would be the best choice for this sentence (reproduced below)?

The diction of the story is also more precise and can be either straightforward or suggestive (43) <u>depending in earnest on the goal of the writer.</u>

a. NO CHANGE
b. depending; in earnest on the goal of the writer.
c. depending, in earnest, on the goal of the writer.
d. the goal of the writer, in earnest, depends on the goal of the writer.

44. Which of the following would be the best choice for this sentence (reproduced below)?

Usually, all news (44) <u>outlets have some sort of bias, it's just a question of how much</u> bias clouds the reporting.

a. NO CHANGE
b. outlets have some sort of bias. Just a question of how much
c. outlets have some sort of bias it can just be a question of how much
d. outlets have some sort of bias, its just a question of how much

Math

1. If $6t + 4 = 16$, what is t?
    a. 1
    b. 2
    c. 3
    d. 4

2. The variable y is directly proportional to x. If $y = 3$ when $x = 5$, then what is y when $x = 20$?
    a. 10
    b. 12
    c. 14
    d. 16

3. A line passes through the point (1, 2) and crosses the y-axis at $y = 1$. Which of the following is an equation for this line?
    a. $y = 2x$
    b. $y = x + 1$
    c. $x + y = 1$
    d. $y = \frac{x}{2} - 2$

4. There are $4x + 1$ treats in each party favor bag. If a total of $60x + 15$ treats are distributed, how many bags are given out?
    a. 15
    b. 16
    c. 20
    d. 22

5. Apples cost $2 each, while oranges cost $3 each. Maria purchased 10 fruits in total and spent $22. How many apples did she buy?
  a. 5
  b. 6
  c. 7
  d. 8

6. What are the polynomial roots of $x^2 + x - 2$?
  a. 1 and -2
  b. -1 and 2
  c. 2 and -2
  d. 9 and 13

7. What is the y-intercept of $y = x^{5/3} + (x - 3)(x + 1)$?
  a. 3.5
  b. 7.6
  c. -3
  d. -15.1

8. $x^4 - 16$ can be simplified to which of the following?
  a. $(x^2 - 4)(x^2 + 4)$
  b. $(x^2 + 4)(x^2 + 4)$
  c. $(x^2 - 4)(x^2 - 4)$
  d. $(x^2 - 2)(x^2 + 4)$

9. $(4x^2y^4)^{\frac{3}{2}}$ can be simplified to which of the following?
  a. $8x^3y^6$
  b. $4x^{\frac{5}{2}}y$
  c. $4xy$
  d. $32x^{\frac{7}{2}}y^{\frac{11}{2}}$

10. If $\sqrt{1 + x} = 4$, what is x?
  a. 10
  b. 15
  c. 20
  d. 25

11. Suppose $\frac{x+2}{x} = 2$. What is x?
  a. -1
  b. 0
  c. 2
  d. 4

163

12. A ball is thrown from the top of a high hill, so that the height of the ball as a function of time is $h(t) = -16t^2 + 4t + 6$, in feet. What is the maximum height of the ball in feet?

    a. 6
    b. 6.25
    c. 6.5
    d. 6.75

13. A rectangle has a length that is 5 feet longer than three times its width. If the perimeter is 90 feet, what is the length in feet?

    a. 10
    b. 20
    c. 25
    d. 35

14. Five students take a test. The scores of the first four students are 80, 85, 75, and 60. If the median score is 80, which of the following could NOT be the score of the fifth student?

    a. 60
    b. 80
    c. 85
    d. 100

15. In an office, there are 50 workers. A total of 60% of the workers are women, and the chances of a woman wearing a skirt is 50%. If no men wear skirts, how many workers are wearing skirts?

    a. 12
    b. 15
    c. 16
    d. 20

16. An arc is intercepted by a central angle of 240°. What is the number of radians of that angle?

    a. $\frac{3\pi}{4}$
    b. $\frac{4\pi}{3}$
    c. $\frac{\pi}{4}$
    d. $\frac{\pi}{3}$

17. A company invests $50,000 in a building where they can produce saws. If the cost of producing one saw is $40, then which function expresses the amount of money the company pays? The variable y is the money paid and x is the number of saws produced.

    a. $y = 50,000x + 40$
    b. $y + 40 = x - 50,000$
    c. $y = 40x - 50,000$
    d. $y = 40x + 50,000$

18. A six-sided die is rolled. What is the probability that the roll is 1 or 2?

a. $\frac{1}{6}$

b. $\frac{1}{4}$

c. $\frac{1}{3}$

d. $\frac{1}{2}$

19. A line passes through the origin and through the point (-3, 4). What is the slope of the line?

a. $-\frac{4}{3}$

b. $-\frac{3}{4}$

c. $\frac{4}{3}$

d. $\frac{3}{4}$

20. What is the equation of a circle whose center is (0, 0) and whole radius is 5?

a. $(x - 5)^2 + (y - 5)^2 = 25$
b. $(x)^2 + (y)^2 = 5$
c. $(x)^2 + (y)^2 = 25$
d. $(x + 5)^2 + (y + 5)^2 = 25$

21.

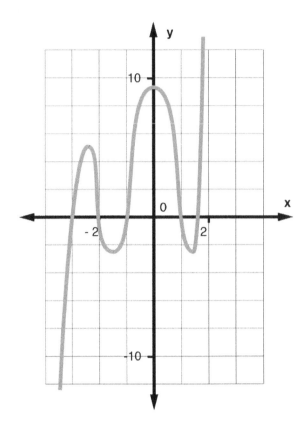

Which of the following functions represents the graph above?

    a. $y = x^5 + 3.5x^4 - 2.5x^2 + 1.5x + 9$
    b. $y = x^5 - 3.5x^4 + 2.5x^2 - 1.5x - 9$
    c. $y = 5x^4 - 2.5x^2 + 1.5x + 9$
    d. $y = -5x^4 - 2.5x^2 + 1.5x + 9$

22. Katie works at a clothing company and sold 192 shirts over the weekend. $\frac{1}{3}$ of the shirts that were sold were patterned, and the rest were solid. Which mathematical expression would calculate the number of solid shirts Katie sold over the weekend?

    a. $192 \times \frac{1}{3}$

    b. $192 \div \frac{1}{3}$

    c. $192 \times (1 - \frac{1}{3})$

    d. $192 \div 3$

23. What is the equation of a circle whose center is (1, 5) and whole radius is 4?

    a. $(x - 1)^2 + (y - 25)^2 = 4$
    b. $(x - 1)^2 + (y - 25)^2 = 16$
    c. $(x + 1)^2 + (y + 5)^2 = 16$
    d. $(x - 1)^2 + (y - 5)^2 = 16$

24. Given the value of a given stock at monthly intervals, which graph should be used to best represent the trend of the stock?
    a. Box plot
    b. Line plot
    c. Line graph
    d. Circle graph

25. What is the probability of randomly picking the winner and runner-up from a race of 4 horses and distinguishing which is the winner?
    a. $\frac{1}{4}$

    b. $\frac{1}{2}$

    c. $\frac{1}{16}$

    d. $\frac{1}{12}$

26. What is the next number in the following series: $1, 3, 6, 10, 15, 21, \dots$?
    a. 26
    b. 27
    c. 28
    d. 29

27. A shipping box has a length of 8 inches, a width of 14 inches, and a height of 4 inches. If all three dimensions are doubled, what is the relationship between the volume of the new box and the volume of the original box?
    a. The volume of the new box is double the volume of the original box.
    b. The volume of the new box is four times as large as the volume of the original box.
    c. The volume of the new box is six times as large as the volume of the original box.
    d. The volume of the new box is eight times as large as the volume of the original box.

28. What is the product of the following expression?
$$(3 + 2i)(5 - 4i).$$
    a. $23 - 2i$
    b. $15 - 8i$
    c. $15 - 8i^2$
    d. $15 - 10i$

29.

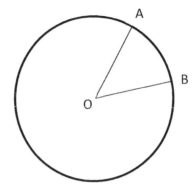

The length of arc $AB = 3\pi$ cm. The length of $\overline{OA} = 12$ cm. What is the degree measure of $\angle AOB$?
  a. 30 degrees
  b. 40 degrees
  c. 45 degrees
  d. 55 degrees

30. How will the following algebraic expression be simplified: $(5x^2 - 3x + 4) - (2x^2 - 7)$?
  a. x^5
  b. $3x^2 - 3x + 11$
  c. $3x^2 - 3x - 3$
  d. $x - 3$

31. The triangle shown below is a right triangle. What's the value of x?

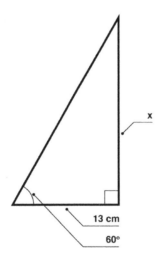

  a. $x = 15.4$
  b. $x = 54.2$
  c. $x = 13$
  d. $x = 22.52$

32. A ball is drawn at random from a ball pit containing 8 red balls, 7 yellow balls, 6 green balls, and 5 purple balls. What's the probability that the ball drawn is yellow?

a. $\frac{1}{26}$

b. $\frac{19}{26}$

c. $\frac{7}{26}$

d. 1

33. If $-3(x + 4) \geq x + 8$, what is the value of x?

a. $x = 4$
b. $x \geq 2$
c. $x \geq -5$
d. $x \leq -5$

34. For a group of 20 men, the median weight is 180 pounds and the range is 30 pounds. If each man gains 10 pounds, which of the following would be true?

a. The median weight will increase, and the range will remain the same.
b. The median weight and range will both remain the same.
c. The median weight will stay the same, and the range will increase.
d. The median weight and range will both increase.

35. If the ordered pair $(-3, -4)$ is reflected over the x-axis, what's the new ordered pair?

a. $(-3, -4)$
b. $(3, -4)$
c. $(3, 4)$
d. $(-3, 4)$

36. If the volume of a sphere is 288π cubic meters, what are the radius and surface area of the same sphere?

a. Radius 6 meters and surface area 144π square meters
b. Radius 36 meters and surface area 144π square meter.
c. Radius 6 meters and surface area 12π square meters
d. Radius 36 meters and surface area 12π square meters

37. Which four-sided shape is always a rectangle?

a. Rhombus
b. Square
c. Parallelogram
d. Quadrilateral

38. Using trigonometric ratios for a right angle, what is the value of the angle whose opposite side is equal to 25 centimeters and whose hypotenuse is equal to 50 centimeters?

a. $15°$
b. $30°$
c. $45°$
d. $90°$

39. What is the length of a chord, whose angle subtended at the center by the chord is 60°, and whose radius is 30 cm?

    a. 5 cm

    b. 10 cm

    c. 30 cm

    d. 20 cm

40. What is the function that forms an equivalent graph to $y = \cos(x)$?

    a. $y = \tan(x)$

    b. $y = \csc(x)$

    c. $y = \sin(x + \frac{\pi}{2})$

    d. $y = \sin(x - \frac{\pi}{2})$

41. A solution needs 5 mL of saline for every 8 mL of medicine given. How much saline is needed for 45 mL of medicine?

    a. $\frac{225}{8}$ mL

    b. 72 mL

    c. 28 mL

    d. $\frac{45}{8}$ mL

42. What's the midpoint of a line segment with endpoints $(-1, 2)$ and $(3, -6)$?

    a. $(1, 2)$

    b. $(1, 0)$

    c. $(-1, 2)$

    d. $(1, -2)$

43. A sample data set contains the following values: 1, 3, 5, 7. What's the standard deviation of the set?

    a. 2.58

    b. 4

    c. 6.23

    d. 1.1

No Calculator Questions

44. An equilateral triangle has a perimeter of 18 feet. If a square whose sides have the same length as one side of the triangle is built, what will be the area of the square?

    a. 6 square feet

    b. 36 square feet

    c. 256 square feet

    d. 1000 square feet

45. If a car can travel 300 miles in 4 hours, how far can it go in an hour and a half?

    a. 100 miles

    b. 112.5 miles

    c. 135.5 miles

    d. 150 miles

46. Ten students take a test. Five students get a 50. Four students get a 70. If the average score is 55, what was the last student's score?
    a. 20
    b. 40
    c. 50
    d. 60

47. A pizzeria owner regularly creates jumbo pizzas, each with a radius of 9 inches. She is mathematically inclined, and wants to know the area of the pizza to purchase the correct boxes and know how much she is feeding her customers. What is the area of the circle, in terms of π, with a radius of 9 inches?
    a. $81 \pi \ in^2$
    b. $18 \pi \ in^2$
    c. $90 \pi \ in^2$
    d. $9 \pi \ in^2$

48. How will the following number be written in standard form:

$$(1 \times 10^4) + (3 \times 10^3) + (7 \times 10^1) + (8 \times 10^0)$$

    a. 137
    b. 13,078
    c. 1,378
    d. 8,731

49. What is the simplified form of the expression $tan\theta \ cos\theta$?
    a. $sin\theta$
    b. 1
    c. $csc\theta$
    d. $\frac{1}{sec\theta}$

50. What is the value of the sum of $\frac{1}{3}$ and $\frac{2}{5}$?
    a. $\frac{3}{8}$

    b. $\frac{11}{15}$

    c. $\frac{11}{30}$

    d. $\frac{4}{5}$

51. If the cosine of $30° = x$, the sine of what angle also equals x?
    a. $30°$
    b. $60°$
    c. $90°$
    d. $120°$

52. If sine of 60° = x, the cosine of what angle also equals x?
    a. 30°
    b. 60°
    c. 90°
    d. 120°

53.

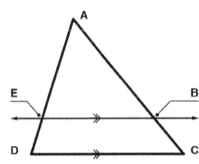

If $\overline{AE} = 4$, $\overline{AB} = 5$, and $\overline{AD} = 5$, what is the length of $\overline{AC}$?

54. $\frac{3}{25} =$

55. 6 is 30% of what number?

56. What is the value of the following expression?

$$\sqrt{8^2 + 6^2}$$

Answer Explanations for Practice Test #1

Reading

1. B: Mr. Button's wife is about to have a baby. The passage begins by giving the reader information about traditional birthing situations. Then, we are told that Mr. and Mrs. Button decide to go against tradition to have their baby in a hospital. The next few passages are dedicated to letting the reader know how Mr. Button dresses and goes to the hospital to welcome his new baby. There is a doctor in this excerpt, as Choice *C* indicates, and Mr. Button does put on clothes, as Choice *D* indicates. However, Mr. Button is not going to the doctor's office nor is he about to go shopping for new clothes.

2. A: The tone of the above passage is nervous and excited. We are told in the fourth paragraph that Mr. Button "arose nervously." We also see him running without caution to the doctor to find out about his wife and baby—this indicates his excitement. We also see him stuttering in a nervous yet excited fashion as he asks the doctor if it's a boy or girl. Though the doctor may seem a bit abrupt at the end, indicating a bit of anger or shame, neither of these choices is the overwhelming tone of the entire passage.

3. C: Dedicated. Mr. Button is dedicated to the task before him. Choice *A*, numbed, Choice *B*, chained, and Choice *D*, moved, all could grammatically fit in the sentence. However, they are not synonyms with *consecrated* like Choice *C* is.

4. D: Giving readers a visual picture of what the doctor is doing. The author describes a visual image— the doctor rubbing his hands together—first and foremost. The author may be trying to make a comment about the profession; however, the author does not "explain the detail of the doctor's profession" as Choice *B* suggests.

5. D: To introduce the setting of the story and its characters. We know we are being introduced to the setting because we are given the year in the very first paragraph along with the season: "one day in the summer of 1860." This is a classic structure of an introduction of the setting. We are also getting a long explanation of Mr. Button, what his work is, who is related to him, and what his life is like in the third paragraph.

6. B: "Talk sense!" is an example of an imperative sentence. An imperative sentence gives a command. The doctor is commanding Mr. Button to talk sense. Choice *A* is an example of an exclamatory sentence, which expresses excitement. Choice *C* is an example of an interrogative sentence—these types of sentences ask questions. Choice *D* is an example of a declarative sentence. This means that the character is simply making a statement.

7. C: The point of view is told in third-person omniscient. We know this because the story starts out with us knowing something that the character does not know: that her husband has died. Mrs. Mallard eventually comes to know this, but we as readers know this information before it is broken to her. In third person limited, Choice *D*, we would only see and know what Mrs. Mallard herself knew, and we would find out the news of her husband's death when she found out the news, not before.

8. A: The way Mrs. Mallard reacted to her husband's death. The irony in this story is called situational irony, which means the situation that takes place is different than what the audience anticipated. At the beginning of the story, we see Mrs. Mallard react with a burst of grief to her husband's death. However, once she's alone, she begins to contemplate her future and says the word "free" over and over. This is quite a different reaction from Mrs. Mallard than what readers expected from the first of the story.

9. B: The word "elusive" most closely means "indefinable." Horrible, Choice A, doesn't quite fit with the tone of the word "subtle" that comes before it. Choice C, "quiet," is more closely related to the word "subtle." Choice D, "joyful," also doesn't quite fit the context here. "Indefinable" is the best option.

10. D: Mrs. Mallard, a newly widowed woman, finds unexpected relief in her husband's death. A summary is a brief explanation of the main point of a story. The story mostly focuses on Mrs. Mallard and her reaction to her husband's death, especially in the room when she's alone and contemplating the present and future. All of the other answer choices except Choice C are briefly mentioned in the story; however, they are not the main focus of the story.

11. D: The interesting thing about this story is that feelings that are confused, joyful, and depressive all play a unique and almost equal part of this story. There is no one right answer here, because the author seems to display all of these emotions through the character of Mrs. Mallard. She displays feelings of depressiveness by her grief at the beginning; then, when she receives feelings of joy, she feels moments of confusion. We as readers cannot help but go through these feelings with the character. Thus, the author creates a tone of depression, joy, and confusion, all in one story.

12. C: The word "tumultuously" most nearly means "violently." Even if you don't know the word "tumultuously," look at the surrounding context to figure it out. The next few sentences we see Mrs. Mallard striving to "beat back" the "thing that was approaching to possess her." We see a fearful and almost violent reaction to the emotion that she's having. Thus, her chest would rise and fall tumultuously, or violently.

13. C: In lines 6 and 7, it is stated that avarice can prevent a man from being necessitously poor, but too timorous, or fearful, to achieve real wealth. According to the passage, avarice does tend to make a person very wealthy. The passage states that oppression, not avarice, is the consequence of wealth. The passage does not state that avarice drives a person's desire to be wealthy.

14. D: Paine believes that the distinction that is beyond a natural or religious reason is between king and subjects. He states that the distinction between good and bad is made in heaven. The distinction between male and female is natural. He does not mention anything about the distinction between humans and animals.

15. A: The passage states that the Heathens were the first to introduce government by kings into the world. The quiet lives of patriarchs came before the Heathens introduced this type of government. It was Christians, not Heathens, who paid divine honors to living kings. Heathens honored deceased kings. Equal rights of nature are mentioned in the paragraph, but not in relation to the Heathens.

16. B: Paine asserts that a monarchy is against the equal rights of nature and cites several parts of scripture that also denounce it. He doesn't say it is against the laws of nature. Because he uses scripture to further his argument, it is not despite scripture that he denounces the monarchy. Paine addresses the law by saying the courts also do not support a monarchical government.

17. A: To be *idolatrous* is to worship idols or heroes, in this case, kings. It is not defined as being deceitful. While idolatry is considered a sin, it is an example of a sin, not a synonym for it. Idolatry may have been considered illegal in some cultures, but it is not a definition for the term.

18. A: The essential meaning of the passage is that the Almighty, God, would disapprove of this type of government. While heaven is mentioned, it is done so to suggest that the monarchical government is irreverent, not that heaven isn't promised. God's disapproval is mentioned, not his punishment. The passage refers to the Jewish monarchy, which required both belief in God and kings.

19. A: Choice *D* can be eliminated because the Salem witch trials aren't even mentioned. While sympathetic to the plight of the accused, the author doesn't demand or urge the reader to demand reparations to the descendants; therefore, Choice *B* can also be ruled out. It's clear that the author's main goal is to educate the reader and shed light on the facts and hidden details behind the case. However, his focus isn't on the occult, but the specific Lancashire case itself. He goes into detail about suspects' histories and ties to Catholicism, revealing how the fears of the English people at the time sealed the fate of the accused witches. Choice *A* is correct.

20. B: It's important to note that these terms may not be an exact analog for *enduring*. However, through knowledge of the definition of *enduring*, as well as the context in which it's used, an appropriate synonym can be found. Plugging "circumstantial" into the passage in place of "enduring" doesn't make sense. Nor does "un-original," this particular case of witchcraft, stand out in history. "Wicked" is very descriptive, but this is an attribute applied to people, not events; therefore, this is an inappropriate choice as well. *Enduring* literally means long lasting, referring to the continued interest in this particular case of witchcraft. Therefore, it's a popular topic of 1600s witch trials, making "popular," Choice *B*, the best choice.

21. D: Choices A and B are irrelevant. The use of quotes lends credibility to the author. However, the presence of quotes alone doesn't necessarily mean that the author has a qualified perspective. What establishes the writer as a reliable voice is that the author's previous writing on the subject has been published before. This qualification greatly establishes the author's credentials as a historical writer, making Choice D the correct answer.

22. B: Choice *A* is incorrect, clearly taking the statement somewhat literally. The remaining three choices appear somewhat interconnected, and though they may be proven at some point later in the article, the focus must remain on the given excerpt. It's very possible that evidence was tampered with or even falsified, but this statement doesn't refer to this. While the author alludes that there may have been evidence tampering and potentially corruption, what the writer is directly saying is that the documentation of the court indicates an elaborate trial. It's clear that exaggerations may have taken place both during the case and in the written account. The reasoning behind this was to gain the attention of the people and even the crown. Choice *B* is the best answer because it not only aligns with the above statement, but ultimately encompasses the potentiality of Choices *C* and D as well.

23. C: Several of these answers could have contributed to the fear and political motivations around the Lancashire witch trials. What this answer's looking for is very specific: political motivations and issues that played a major role in the case. Choice C clearly outlines the public fears of the time. It also describes how the government can use this fear to weed out and eliminate traces of Catholicism (and witchcraft too). Catholicism and witchcraft were seen as dangerous and undermining to English Protestantism and governance. Choice D can be eliminated; while this information may have some truth and is certainly consistent with the general fear of witchcraft, the details about Lancashire's ancient history aren't mentioned in the text. Choice A is true but not necessarily political in nature. Choice B is very promising, though not outright mentioned.

24. D: The best evidence comes from Alizon herself. The text mentions that she confessed to bewitching John Law, thinking that she did him harm. From here she names her grandmother, who she believes corrupted her. Choice B can be ruled out; spectral evidence isn't mentioned. The case draws on knowledge of superstition of witchcraft, but this in itself can't be considered evidence, so Choice A is incorrect. Choice C isn't evidence in a modern sense; rumors have no weight in court and therefore are not evidence. While this is used as evidence to some degree, this still isn't the best evidence against Alizon and the witches.

25. C: It's very clear that the author is chronicling the development of museums, but is there a defined viewpoint that he is trying to impress upon the reader? No. The author presents facts, and even inferences, but he is clearly not trying to claim something new. Instead, he's simply providing details about the rise of museums by describing the first examples of people collecting and displaying items. This makes Choice *A* and Choice *D* incorrect. Also, the 1850s weren't even mentioned in this excerpt. The author doesn't insert himself into the piece; there's no use of "I." There also isn't any mention of his own experiences. This also means that Choice *B* is incorrect. Clearly, the author is documenting the development of museums by focusing on collections that, while not conforming to modern museums, would give rise to the idea later on. Choice *C* is correct.

26. B: Generally, all the questions seem to hold some truth, but the goal is to find the answer that best explains the excerpt. Choice *C* can be eliminated first. It doesn't address the statement directly; besides, there are traveling exhibits. Choice *D* is very true, but this doesn't provide the reasoning behind the statement. Choice *A* is very strong, but Choice *B* is actually stronger. Choice *B* addresses the fact that a stable society enables the museum to be maintained and visitors to actively visit the museum. Thus, Choice *B* is correct.

27. D: Recall that the text illustrates that initially it was primarily royalty that owned collections and displayed them, but this didn't necessarily mean that these artifacts made up a museum. This eliminates Choice *A*. The author goes on to reflect on how the great institute of Alexandria wouldn't really be considered a museum because "the name museum in that institution was applied to a portion set apart for the study of sciences, and indicated rather a place of study than one for exhibition of objects." While a museum is a place of study, it's clear that the modern museum is primarily geared for exhibition. With this in mind, Choice *B* simply falls short and can be eliminated. The answer that sums up the core criteria for a modern museum is Choice *D*. The purpose of a modern museum is not just to house artifacts but to display them for the public. With this in mind, Choice *C* falls short by not addressing the core role of a museum.

28. A: Choice *C* can be eliminated first; this answer makes no sense. The term "virtues" is modified by "preservative." In other words, the phrase "preservative virtues" is talking about a substance's potential to provide preservation. Preservation is not an act of correction; this doesn't make sense, so Choice *B* can be eliminated. Choice *D* is very promising, but "virtues" isn't referring to the capability of a substance but rather the good qualities within the alcohol. In this case, "virtues" is used as a noun that means specific qualities. Therefore, the most accurate alternative would be "properties," Choice *A*. *Properties* is a noun that means distinctive attributes or qualities.

29. C: Avoid choosing answers that seem exaggerated or subjective. It's well known that Aristotle is an accomplished philosopher and natural history researcher, but this passage doesn't quantify his

expertise. While we can infer that Aristotle is highly adept to be highly favored by Alexander, Choice *A* is an opinion that isn't directly stated. Choice *B* may be true, but this is too broad to gauge from this single excerpt. Choice *D* could be possible but note how Alexander is gathering the material for Aristotle; the text doesn't hint to additional motives for a museum. The best answer is Choice *C;* if Aristotle has an interest in, or at least intends to study natural artifacts, it stands to reason that others may be actively researching natural history.

30. B: Referring back to the essay, it's clear that this information would fit the best in the fourth paragraph. This is where natural curiosities are first mentioned. However, it introduces the idea of misidentifying the natural curiosities too soon and does not allow the author to actually introduce the natural curiosities and their significance. Inserting this information before the final sentence allows the reader to read about actual examples of ancient natural oddities. This new information would then enable the readers to have a full-circle understanding of the items, showing first ancient interpretations and then modern revelations. Choice *B* is the best answer.

31. C: *Extraneous* most nearly means *superfluous,* or *trivial.* Choice *A, indispensable,* is incorrect because it means the opposite of *extraneous.* Choice *B, bewildering,* means *confusing* and is not relevant to the context of the sentence. Finally, Choice *D* is wrong because although the prefix of the word is the same, *ex-,* the word *exuberant* means *elated* or *enthusiastic,* and is irrelevant to the context of the sentence.

32. A: The author's purpose is to bring to light an alternative view on human perception by examining the role of technology in human understanding. This is a challenging question because the author's purpose is somewhat open-ended. The author concludes by stating that the questions regarding human perception and observation can be approached from many angles. Thus, the author does not seem to be attempting to prove one thing or another. Choice *B* is incorrect because we cannot know for certain whether the electron experiment is the latest discovery in astroparticle physics because no date is given. Choice *C* is a broad generalization that does not reflect accurately on the writer's views. While the author does appear to reflect on opposing views of human understanding (Choice *D*), the best answer is Choice *A.*

33. C: It presents a problem, explains the details of that problem, and then ends with more inquiry. The beginning of this paragraph literally "presents a conundrum," explains the problem of partial understanding, and then ends with more questions, or inquiry. There is no solution offered in this paragraph, making Choices *A* and *B* incorrect. Choice *D* is incorrect because the paragraph does not begin with a definition.

34. D: Looking back in the text, the author describes that classical philosophy holds that understanding can be reached by careful observation. This will not work if they are overly invested or biased in their pursuit. Choices *A, B,* and *C* are in no way related and are completely unnecessary. A specific theory is not necessary to understanding, according to classical philosophy mentioned by the author.

35. B: The electrons passed through both holes and then onto the plate. Choices *A* and *C* are wrong because such movement is not mentioned at all in the text. In the passage the author says that electrons that were physically observed appeared to pass through one hole or another. Remember, the electrons that were observed doing this were described as acting like particles. Therefore, Choice *D* is wrong. Recall that the plate actually recorded electrons passing through both holes simultaneously and hitting the plate. This behavior, the electron activity that wasn't seen by humans, was characteristic of waves. Thus, Choice *B* is the right answer.

36. C: The author mentions "gravity" to demonstrate an example of natural phenomena humans discovered and understood without the use of tools or machines. Choice *A* mirrors the language in the beginning of the paragraph but is incorrect in its intent. Choice *B* is incorrect; the paragraph mentions nothing of "not knowing the true nature of gravity." Choice *D* is incorrect as well. There is no mention of an "alternative solution" in this paragraph.

37. D: If unfamiliar with this term, plugging each term into the sentence in place of "elucidation" will help rule out answers that don't make sense. Using this method, Choices *B* and *C* can be ruled out because they don't fit the sentence. "Definition" looks promising, but this term isn't as comprehensive an explanation. After all, the author is seeking to uncover why myths are the way they are—the reason why there are so many shared concepts. "Rank" is looking for an explanation for why myths appear as they do. In fact, *explanation* is a synonym for *elucidation*. Choice *D* is correct.

38. B: The title and general focus of the passage reveal a lot about the motivations and interests of the writer. This is a study in psychology, so the author's main area of interest is the study of how humans think and process the world. Choice *A* can be eliminated easily. While Rank was in fact a contemporary of Jung, this information is neither relevant nor even mentioned in the text. The other answers are very compelling, but in this case, one must look at what the author already sees in mythology. The reach and similarities in various mythologies suggest a common source—a common humanity. While he doesn't explicitly want to use this knowledge to develop new analysis strategies and therapies (Choice *D*), this does seem to hint to a common human mind-set that would be crucial for his work. Thus, Choice *B* is the best answer. Also, Choice *C* is too narrow and of less interest to the author.

39. A: The key to understanding this theory lies in "the existence of elemental ideas, so that the unanimity of the myths is a necessary sequence of the uniform disposition of the human mind and the manner of its manifestation." In other words, elemental ideas are core questions or ideas that apply to everyone. Choice *B* is compelling, but the idea of winter, and cold climate, doesn't necessarily apply to cultures in climates with mild winters, like in Africa. Choice *C* doesn't offer a lot to consider. Choice *D* is focused on Germanic and Nordic peoples. The best answer will address overarching ideas and concepts that exist throughout the world cultures: elemental ideas. The idea that best exemplifies the Idea of the People theory is Choice *A* because it hints to innate human curiosity and the drive to define the world. Choice *C* can then be ruled out entirely.

40. C: Choice *A* is incorrect; these are not scientific explanations. Choice *B* is incorrect; these are appropriately labeled as "theories," and there is a reason behind this choice. Choice *D* is incorrect and actually a major clue. The author doesn't say that these are competing ideas; rather, he directly says they don't conflict. In truth, all these theories could have a basis in historical fact. However, these can't be proven by physical means. It would be virtually impossible to track, and therefore prove, the exact reason why mythology resonates throughout various cultures and its exact origin. There simply isn't physical evidence. Therefore, these ideas are called *theories* because, while they hold probable truth, they cannot be tested or proven. Choice *C* is correct.

41. D: This can be a very deceptive description, even pessimistic sounding if not considered with the context of the rest of the passage. Choice *A* is incorrect; this takes too rigid a stance on the lines. The quote doesn't condemn people to unoriginality, merely that ideas can be used over and over again, even adapted or copied. Choice *B* looks compelling but doesn't take the rest of the reading into account. Choice *C* is very tempting; it uses similar language to the selection, but note the use of "sundered." Nowhere in the text (both the main passage and the selected quote) is there a mention of sundering or a breakaway from a single culture, but rather the spreading and sharing of ideas. Again, there is also this

hint of a lack of originality that isn't necessarily true. Choice *D*, however, is a phrase that perfectly ties the presented quote with the rest of the passage, a kind of bridge that adds another layer to mythology's origins.

42. B: Choice *A* is compelling, but this statement relates more to the migration theory of spreading mythology. It's a little too broad and less focused on community relations. Choice *D* is compelling and might be the best answer without the presence of Choice *B*. Choice *B* clearly displays what the "explanation by original community idea" focuses on. Note how this theory is centered on a central community (or communities) that expands into different areas. Another key point is that the stories "continued to grow while retaining the common primary traits." In other words, the stories might have grown or changed in some ways, but they still retained core aspects in the narrative. This theory is best exemplified in Choice *B*.

43.B: The author specifically states that the use of objects or tools (extraneous substances) by animals is imitative and not done without human influence. The focus here is on the use of objects to advance an independent desire. Discerning this, Choices *A* and *D* can be eliminated; these do not involve tools or objects. This leaves Choices *B* and *C*. Looking closely at Choice *C*, it's clear that while the monkey used an object, and even used it correctly, this action was done mimicking the human in the video, which makes this answer incorrect. Choice *B*, however, is correct. The fact that wild animals use tools without contact with humans means that they did this independent of human influence and instruction.

44. A: While not defining the term, the author provides a lot of context with which to piece together the meaning of *quadrumana*. From the quotes, it's clear that primates are being discussed. Specifically, the term will indicate their ability to hold on to things. With this in mind, common sense/knowledge comes into play. The first part of the word is very recognizable: "quad." This, of course, is the Latin root for *four*. Another Latin root in the term is *mana*, which refers to hands—think of the term *manual*. Together, the word literally means "four-handed," which makes sense because the text deals with animals that have the potential to hold and manipulate substances. Choice *A* is correct.

45. C: This tricky answer can have a couple of different meanings, which can be lost in the complicated writing style the author uses. In the second paragraph, Thompson writes: "We must conclude, further, that the use of tools and weapons, even preceding the intelligent conformation of them for definite purposes, marked the differentiation of primitive man from the animal branch, and accompanied indeed was the cause of the psychic emergence." Here, the author states clearly that the use of tools and weapons came before the actual use of the objects for intelligent and definite purposes. This is a major clue. The author follows up this statement with: "The moment that the primitive man-ape employed extraneous substances intelligently, with a purpose, he ceased to be an ape and became a man. We must believe that the use of tools and the psychic emergence were coincident and interdependent." This statement puts the psychic emergence on a similar level but not necessarily congruent with free will. This was the consciousness of the possibility of shaping reality that enabled humans to choose what to do. Knowing this, some answers can be easily eliminated. Choices *A* and *B* can be ruled out because their contents aren't addressed at all. Choices *C* and *D* are very close, but remember what was just examined: The psychic emergence isn't necessarily free will, but the ability to process the possibility of using natural resources. Thus, choice *C* is the best answer.

46. B: Before going to the answers, it's important to consider what the original quote means. The original translation of *cogito ergo sum* is "I think; therefore, I am." The author modifies this translation to "I think; therefore I exist." Both translations refer to the ability to self-identify. Essentially, because people can think, they can define their own being and can alter themselves to become what they want.

The fact that people know they exist means in fact they exist in the world. To answer this question, consider how this relates to the use of tools/weapons. The author argues that when our ape ancestors began using resources, they then began to develop uses and designs to advance their own survival. This in turn established their self-awareness and led to higher cognition. The answer here that best combines both ideas is Choice *B*.

47. D: A scepter is a physical object or ornament that symbolizes both sovereignty and authority; it's a literal object that symbolizes power. The author is saying that the tool itself was the device from which man was able to become independent and shape nature to their advantage. With this in mind, it's clear that Choice *D* correctly captures this clever use of symbolism and relates it back to the whole text.

Writing

1. C: Choice *C* correctly uses *from* to describe the fact that dogs are related to wolves. The word *through* is incorrectly used here, so Choice *A* is incorrect. Choice *B* makes no sense. Choice *D* unnecessarily changes the verb tense in addition to incorrectly using *through*.

2. B: Choice *B* is correct because the Oxford comma is applied, clearly separating the specific terms. Choice *A* lacks this clarity. Choice *C* is correct but too wordy since commas can be easily applied. Choice *D* doesn't flow with the sentence's structure.

3. D: Choice *D* correctly uses the question mark, fixing the sentence's main issue. Thus, Choice *A* is incorrect because questions do not end with periods. Choice *B*, although correctly written, changes the meaning of the original sentence. Choice *C* is incorrect because it completely changes the direction of the sentence, disrupts the flow of the paragraph, and lacks the crucial question mark.

4. A: Choice *A* is correct since there are no errors in the sentence. Choices *B* and *C* both have extraneous commas, disrupting the flow of the sentence. Choice *D* unnecessarily rearranges the sentence.

5. D: Choice *D* is correct because the commas serve to distinguish that *artificial selection* is just another term for *selective breeding* before the sentence continues. The structure is preserved, and the sentence can flow with more clarity. Choice *A* is incorrect because the sentence needs commas to avoid being a run-on. Choice *B* is close but still lacks the required comma after *selection*, so this is incorrect. Choice *C* is incorrect because the comma to set off the aside should be placed after *breeding* instead of *called*.

6. B: Choice *B* is correct because the sentence is talking about a continuing process. Therefore, the best modification is to add the word *to* in front of *increase*. Choice *A* is incorrect because this modifier is missing. Choice *C* is incorrect because with the additional comma, the present tense of *increase* is inappropriate. Choice *D* makes more sense, but the tense is still not the best to use.

7. A: The sentence has no errors, so Choice *A* is correct. Choice *B* is incorrect because it adds an unnecessary comma. Choice *C* is incorrect because *advantage* should not be plural in this sentence without the removal of the singular *an*. Choice *D* is very tempting. While this would make the sentence more concise, this would ultimately alter the context of the sentence, which would be incorrect.

8. C: Choice *C* correctly uses *on to*, describing the way genes are passed generationally. The use of *into* is inappropriate for this context, which makes Choice *A* incorrect. Choice *B* is close, but *onto* refers to something being placed on a surface. Choice *D* doesn't make logical sense.

9. D: Choice *D* is correct, since only proper names should be capitalized. Because the name of a dog breed is not a proper name, Choice *A* is incorrect. In terms of punctuation, only one comma after *example* is needed, so Choices *B* and *C* are incorrect.

10. D: Choice *D* is the correct answer because "rather" acts as an interrupting word here and thus should be separated by commas. Choices B and C use commas unwisely, breaking the flow of the sentence.

11. B: Since the sentence can stand on its own without *Usually*, separating it from the rest of the sentence with a comma is correct. Choice *A* needs the comma after *Usually*, while Choice *C* uses commas incorrectly. Choice *D* is tempting but changing *turn* to past tense goes against the rest of the paragraph.

12. A: In Choice *A*, the dependent clause *Sometimes in particularly dull seminars* is seamlessly attached with a single comma after *seminars*. Choice *B* contains too many commas. Choice *C* does not correctly combine the dependent clause with the independent clause. Choice *D* introduces too many unnecessary commas.

13. D: Choice *D* rearranges the sentence to be more direct and straightforward, so it is correct. Choice *A* needs a comma after *on*. Choice *B* introduces unnecessary commas. Choice *C* creates an incomplete sentence, since *Because I wasn't invested in what was going on* is a dependent clause.

14. C: Choice *C* is fluid and direct, making it the best revision. Choice *A* is incorrect because the construction is awkward and lacks parallel structure. Choice *B* is clearly incorrect because of the unnecessary comma and period. Choice *D* is close, but its sequence is still awkward and overly complicated.

15. B: Choice *B* correctly adds a comma after *person* and cuts out the extraneous writing, making the sentence more streamlined. Choice *A* is poorly constructed, lacking proper grammar to connect the sections of the sentence correctly. Choice *C* inserts an unnecessary semicolon and doesn't enable this section to flow well with the rest of the sentence. Choice *D* is better but still unnecessarily long.

16. D: This sentence, though short, is a complete sentence. The only thing the sentence needs is an em-dash after "Easy." In this sentence the em-dash works to add emphasis to the word "Easy" and also acts in place of a colon, but in a less formal way. Therefore, Choice *D* is correct. Choices *A* and *B* lack the crucial comma, while Choice *C* unnecessarily breaks the sentence apart.

17. C: Choice *C* successfully fixes the construction of the sentence, changing *drawing* into *to draw*. Keeping the original sentence disrupts the flow, so Choice *A* is incorrect. Choice *B*'s use of *which* offsets the whole sentence. Choice *D* is incorrect because it unnecessarily expands the sentence content and makes it more confusing.

18. B: Choice *B* fixes the homophone issue. Because the author is talking about people, *their* must be used instead of *there*. This revision also appropriately uses the Oxford comma, separating and distinguishing *lives, world,* and *future*. Choice *A* uses the wrong homophone and is missing commas. Choice *C* neglects to fix these problems and unnecessarily changes the tense of *applies*. Choice *D* fixes the homophone but fails to properly separate *world* and *future*.

19. C: Choice *C* is correct because it fixes the core issue with this sentence: the singular *has* should not describe the plural *scientists*. Thus, Choice *A* is incorrect. Choices *B* and *D* add unnecessary commas.

20. D: Choice *D* correctly conveys the writer's intention of asking if, or *whether*, early perceptions of dinosaurs are still influencing people. Choice *A* makes no sense as worded. Choice *B* is better, but *how* doesn't coincide with the context. Choice *C* adds unnecessary commas.

21. A: Choice *A* is correct, as the sentence does not require modification. Choices *B* and *C* implement extra punctuation unnecessarily, disrupting the flow of the sentence. Choice *D* incorrectly adds a comma in an awkward location.

22. B: Choice *B* is the strongest revision, as adding *to explore* is very effective in both shortening the sentence and maintaining, even enhancing, the point of the writer. To explore is to seek understanding in order to gain knowledge and insight, which coincides with the focus of the overall sentence. Choice *A* is not technically incorrect, but it is overcomplicated. Choice *C* is a decent revision, but the sentence could still be more condensed and sharpened. Choice *D* fails to make the sentence more concise and inserts unnecessary commas.

23. D: Choice *D* correctly applies a semicolon to introduce a new line of thought while remaining in a single sentence. The comma after *however* is also appropriately placed. Choice *A* is a run-on sentence. Choice *B* is also incorrect because the single comma is not enough to fix the sentence. Choice *C* adds commas around *uncertain* which are unnecessary.

24. B: Choice *B* not only fixes the homophone issue from *its*, which is possessive, to *it's*, which is a contraction of *it is*, but also streamlines the sentence by adding a comma and eliminating *and*. Choice *A* is incorrect because of these errors. Choices *C* and *D* only fix the homophone issue.

25. A: Choice *A* is correct, as the sentence is fine the way it is. Choices *B* and *C* add unnecessary commas, while Choice *D* uses the possessive *its* instead of the contraction *it's*.

26. C: Choice *C* is correct because the phrase *even likely* is flanked by commas, creating a kind of aside, which allows the reader to see this separate thought while acknowledging it as part of the overall sentence and subject at hand. Choice *A* is incorrect because it seems to ramble after *even* due to a missing comma after *likely*. Choice *B* is better but inserting a comma after *that* warps the flow of the writing. Choice *D* is incorrect because there must be a comma after *plausible*.

27. D: Choice *D* strengthens the overall sentence structure while condensing the words. This makes the subject of the sentence, and the emphasis of the writer, much clearer to the reader. Thus, while Choice *A* is technically correct, the language is choppy and over-complicated. Choice *B* is better but lacks the reference to a specific image of dinosaurs. Choice *C* introduces unnecessary commas.

28. B: Choice *B* correctly joins the two independent clauses. Choice *A* is decent, but "that would be" is too verbose for the sentence. Choice *C* incorrectly changes the semicolon to a comma. Choice *D* splits the clauses effectively but is not concise enough.

29. A: Choice *A* is correct, as the original sentence has no error. Choices *B* and *C* employ unnecessary semicolons and commas. Choice *D* would be an ideal revision, but it lacks the comma after *Ransom* that would enable the sentence structure to flow.

30. D: By reorganizing the sentence, the context becomes clearer with Choice *D*. Choice *A* has an awkward sentence structure. Choice *B* offers a revision that doesn't correspond well with the original sentence's intent. Choice *C* cuts out too much of the original content, losing the full meaning.

31. C: Choice *C* fixes the disagreement between the singular *this* and the plural *viewpoints*. Choice *A*, therefore, is incorrect. Choice *B* introduces an unnecessary comma. In Choice *D*, *those* agrees with *viewpoints*, but neither agrees with *distinguishes*.

32. A: Choice *A* is direct and clear, without any punctuation errors. Choice *B* is well-written but too wordy. Choice *C* adds an unnecessary comma. Choice *D* is also well-written but much less concise than Choice *A*.

33. D: Choice *D* rearranges the sentence to improve clarity and impact, with *tempting* directly describing *idea*. On its own, Choice *A* is a run-on. Choice *B* is better because it separates the clauses, but it keeps an unnecessary comma. Choice *C* is also an improvement but still a run-on.

34. B: Choice *B* is the best answer simply because the sentence makes it clear that Un-man takes over and possesses Weston. In Choice *A*, these events sounded like two different things, instead of an action and result. Choices *C* and *D* make this relationship clearer, but the revisions don't flow very well grammatically.

35. D: Changing the phrase *after this* to *then* makes the sentence less complicated and captures the writer's intent, making Choice *D* correct. Choice *A* is awkwardly constructed. Choices *B* and *C* misuse their commas and do not adequately improve the clarity.

36. B: By starting a new sentence, the run-on issue is eliminated, and a new line of reasoning can be seamlessly introduced, making Choice *B* correct. Choice *A* is thus incorrect. While Choice *C* fixes the run-on via a semicolon, a comma is still needed after *this*. Choice *D* contains a comma splice. The independent clauses must be separated by more than just a comma, even with the rearrangement of the second half of the sentence.

37. C: Choice *C* condenses the original sentence while being more active in communicating the emphasis on changing times/media that the author is going for, so it is correct. Choice *A* is clunky because it lacks a comma after *today* to successfully transition into the second half of the sentence. Choice *B* inserts unnecessary commas. Choice *D* is a good revision of the underlined section, but not only does it not fully capture the original meaning, it also does not flow into the rest of the sentence.

38. B: Choice *B* clearly illustrates the author's point, with a well-placed semicolon that breaks the sentence into clearer, more readable sections. Choice *A* lacks punctuation. Choice *C* is incorrect because the period inserted after *question* forms an incomplete sentence. Choice *D* is a very good revision but does not make the author's point clearer than the original.

39. A: Choice *A* is correct: while the sentence seems long, it actually doesn't require any commas. The conjunction "that" successfully combines the two parts of the sentence without the need for additional punctuation. Choices *B* and *C* insert commas unnecessarily, incorrectly breaking up the flow of the sentence. Choice *D* alters the meaning of the original text by creating a new sentence, which is only a fragment.

40. C: Choice *C* correctly replaces *for* with *to*, the correct preposition for the selected area. Choice *A* is not the answer because of this incorrect preposition. Choice *B* is unnecessarily long and disrupts the original sentence structure. Choice *D* is also too wordy and lacks parallel structure.

41. D: Choice *D* is the answer because it inserts the correct punctuation to fix the sentence, linking the dependent and independent clauses. Choice *A* is therefore incorrect. Choice *B* is also incorrect since this revision only adds content to the sentence while lacking grammatical precision. Choice C overdoes the punctuation; only a comma is needed, not a semicolon.

42. B: Choice *B* correctly separates the section into two sentences and changes the word order to make the second part clearer. Choice *A* is incorrect because it is a run-on. Choice *C* adds an extraneous comma, while Choice *D* makes the run-on worse and does not coincide with the overall structure of the sentence.

43. C: Choice *C* is the best answer because of how the commas are used to flank *in earnest*. This distinguishes the side thought (*in earnest*) from the rest of the sentence. Choice *A* needs punctuation. Choice *B* inserts a semicolon in a spot that doesn't make sense, resulting in a fragmented sentence and lost meaning. Choice *D* is unnecessarily elaborate and leads to a run-on.

44. A: Choice *A* is correct because the sentence contains no errors. The comma after *bias* successfully links the two halves of the sentence, and the use of *it's* is correct as a contraction of *it is*. Choice *B* creates a sentence fragment, while Choice *C* creates a run-on. Choice *D* incorrectly changes *it's* to *its*.

Math

1. B: First, subtract 4 from each side. This yields $6t = 12$. Now, divide both sides by 6 to obtain $t = 2$.

2. B: To be directly proportional means that $y = mx$. If x is changed from 5 to 20, the value of x is multiplied by 4. Applying the same rule to the y-value, also multiply the value of y by 4. Therefore, $y = 12$.

3. B: From the slope-intercept form, $y = mx + b$, it is known that b is the y-intercept, which is 1. Compute the slope as $\frac{2-1}{1-0} = 1$, so the equation should be $y = x + 1$.

4. A: Each bag contributes $4x + 1$ treats. The total treats will be in the form $4nx + n$ where n is the total number of bags. The total is in the form $60x + 15$, from which it is known $n = 15$.

5. D: Let a be the number of apples and o the number of oranges. Then, the total cost is $2a + 3o = 22$, while it also known that $a + o = 10$. Using the knowledge of systems of equations, cancel the o variables by multiplying the second equation by -3. This makes the equation $-3a - 3o = -30$. Adding this to the first equation, the b values cancel to get $-a = -8$, which simplifies to $a = 8$.

6. A: Finding the roots means finding the values of x when y is zero. The quadratic formula could be used, but in this case it is possible to factor by hand, since the numbers -1 and 2 add to 1 and multiply to -2. So, factor $x^2 + x - 2 = (x - 1)(x + 2) = 0$, then set each factor equal to zero. Solving for each value gives the values $x = 1$ and $x = -2$.

7. C: To find the y-intercept, substitute zero for x, which gives us:

$$y = 0^{\frac{5}{3}} + (0 - 3)(0 + 1)$$

$$0 + (-3)(1) = -3$$

8. A: This has the form $t^2 - y^2$, with $t = x^2$ and $y = 4$. It's also known that $t^2 - y^2 = (t + y)(t - y)$, and substituting the values for t and y into the right-hand side gives:

$$(x^2 - 4)(x^2 + 4)$$

9. A: Simplify this to:

$$(4x^2y^4)^{\frac{3}{2}} = 4^{\frac{3}{2}}(x^2)^{\frac{3}{2}}(y^4)^{\frac{3}{2}}$$

Now:

$$4^{\frac{3}{2}} = (\sqrt{4})^3 = 2^3 = 8$$

For the other, recall that the exponents must be multiplied, so this yields:

$$8x^{2 \cdot \frac{3}{2}}y^{4 \cdot \frac{3}{2}} = 8x^3y^6$$

10. B: Start by squaring both sides to get $1 + x = 16$. Then subtract 1 from both sides to get $x = 15$.

11. C: Multiply both sides by x to get $x + 2 = 2x$, which simplifies to $-x = -2$, or $x = 2$.

12. B: The independent variable's coordinate at the vertex of a parabola (which is the highest point, when the coefficient of the squared independent variable is negative) is given by $x = -\frac{b}{2a}$. Substitute and solve for x to get:

$$x = -\frac{4}{2(-16)} = \frac{1}{8}$$

Using this value of x, the maximum height of the ball (y), can be calculated. Substituting x into the equation yields:

$$h(t) = -16\frac{1}{8}^2 + 4\frac{1}{8} + 6 = 6.25$$

13. D: Denote the width as w and the length as l. Then, $l = 3w + 5$. The perimeter is $2w + 2l = 90$. Substituting the first expression for l into the second equation yields:

$$2(3w + 5) + 2w = 90$$

Which then simplifies to:

$$8w = 80$$

Which also equals:

$$w = 10$$

Putting this into the first equation, it yields:

$$l = 3(10) + 5 = 35$$

14. A: Lining up the given scores provides the following list: 60, 75, 80, 85, and one unknown. Because the median needs to be 80, it means 80 must be the middle data point out of these five. Therefore, the unknown data point must be the fourth or fifth data point, meaning it must be greater than or equal to 80. The only answer that fails to meet this condition is 60.

15. B: If 60% of 50 workers are women, then there are 30 women working in the office. If half of them are wearing skirts, then that means 15 women wear skirts. Since none of the men wear skirts, this means there are 15 people wearing skirts.

16. B: When you simplify $(\frac{4\pi}{3}) \times (\frac{\pi}{180})$, you get $\frac{4\pi}{3}$. Choice A is not the correct answer because it is the reciprocal of $\frac{4\pi}{3}$. Choice C is not the correct answer because it is not the correct reduction of the fraction. Choice D is not the correct answer because it is not the correct reduction of the fraction.

17. D: For manufacturing costs, there is a linear relationship between the cost to the company and the number produced, with a y-intercept given by the base cost of acquiring the means of production, and a slope given by the cost to produce one unit. In this case, that base cost is $50,000, while the cost per unit is $40. So:

$$y = 40x + 50{,}000$$

18. C: A die has an equal chance for each outcome. Since it has six sides, each outcome has a probability of $\frac{1}{6}$. The chance of a 1 or a 2 is therefore:

$$\frac{1}{6} + \frac{1}{6} = \frac{1}{3}$$

19. A: The slope is given by:

$$m = \frac{y_2 - y_1}{x_2 - x_1} = \frac{0 - 4}{0 - (-3)} = -\frac{4}{3}$$

20. C: Nothing is added to x and y since the center is 0 and 5^2 is 25. Choice A is not the correct answer because you do not subtract the radius from x and y. Choice B is not the correct answer because you must square the radius on the right side of the equation. Choice D is not the correct answer because you do not add the radius to x and y in the equation.

21. A: The graph contains four turning points (where the curve changes from rising to falling or vice versa). This indicates that the degree of the function (highest exponent for the variable) is 5, eliminating Choices C and D. The y-intercepts of the functions can be determined by substituting 0 for x and finding the value of y. The function for Choice A has a y-intercept of 9, and the function for Choice B has a y-intercept of -9. Therefore, Choice B is eliminated.

22. C: $\frac{1}{3}$ of the shirts sold were patterned. Therefore, $1 - \frac{1}{3} = \frac{2}{3}$ of the shirts sold were solid. Anytime "of" a quantity appears in a word problem, multiplication should be used. Therefore:

$$192 \times \frac{2}{3} = \frac{192 \times 2}{3} = \frac{384}{3} = 128 \text{ solid shirts were sold}$$

The entire expression is:

$$192 \times \left(1 - \frac{1}{3}\right)$$

23. D: Subtract the center from the x and y values of the equation and square the radius on the right side of the equation. Choice A is not the correct answer because you need to square the radius of the equation. Choice B is not the correct answer because you do not square the centers of the equation. Choice C is not the correct answer because you need to subtract (not add) the centers of the equation.

24. C: Line graph. The scenario involves data consisting of two variables, month and stock value. Box plots display data consisting of values for one variable. Therefore, a box plot is not an appropriate choice. Both line plots and circle graphs are used to display frequencies within categorical data. Neither can be used for the given scenario. Line graphs display two numerical variables on a coordinate grid and show trends among the variables.

25. D: $\frac{1}{12}$. The probability of picking the winner of the race is:

$$\frac{1}{4}\left(\frac{number\ of\ favorable\ outcomes}{number\ of\ total\ outcomes}\right)$$

Assuming the winner was picked on the first selection, three horses remain from which to choose the runner-up (these are dependent events). Therefore, the probability of picking the runner-up is $\frac{1}{3}$. To determine the probability of multiple events, the probability of each event is multiplied:

$$\frac{1}{4} \times \frac{1}{3} = \frac{1}{12}.$$

26. C: Each number in the sequence is adding one more than the difference between the previous two. For example, $10 - 6 = 4, 4 + 1 = 5$. Therefore, the next number after 10 is $10 + 5 = 15$. Going forward, $21 - 15 = 6, 6 + 1 = 7$. The next number is $21 + 7 = 28$. Therefore, the difference between numbers is the set of whole numbers starting at 2: 2, 3, 4, 5, 6, 7....

27. D: The formula for finding the volume of a rectangular prism is $V = l \times w \times h$ where l is the length, w is the width, and h is the height. The volume of the original box is calculated:

$$V = 8 \times 14 \times 4 = 448 \text{ in}^3$$

The volume of the new box is calculated:

$$V = 16 \times 28 \times 8 = 3584 \text{ in}^3$$

The volume of the new box divided by the volume of the old box equals 8.

28. A: The notation i stands for an imaginary number. The value of i is equal to $\sqrt{-1}$. When performing calculations with imaginary numbers, treat i as a variable, and simplify when possible. Multiplying the binomials by the FOIL method produces:

$$15 - 12i + 10i - 8i^2$$

Combining like terms yields $15 - 2i - 8i^2$. Since:

$$i = \sqrt{-1}, i^2 = (\sqrt{-1})^2 = -1$$

Therefore, substitute -1 for i^2:

$$15 - 2i - 8(-1)$$

Simplifying results in:

$$15 - 2i + 8 \rightarrow 23 - 2i$$

29. C: The formula to find arc length is $s = \theta r$ where s is the arc length, θ is the radian measure of the central angle, and r is the radius of the circle. Substituting the given information produces

3π cm $= \theta 12$ cm. Solving for θ yields $\theta = \frac{\pi}{4}$. To convert from radian to degrees, multiply the radian measure by $\frac{180}{\pi}$:

$$\frac{\pi}{4} \times \frac{180}{\pi} = 45^\circ$$

30. B: $3x^2 - 3x + 11$. By distributing the implied one in front of the first set of parentheses and the -1 in front of the second set of parentheses, the parenthesis can be eliminated:

$$1(5x^2 - 3x + 4) - 1(2x^2 - 7)$$

$$5x^2 - 3x + 4 - 2x^2 + 7$$

Next, like terms (same variables with same exponents) are combined by adding the coefficients and keeping the variables and their powers the same:

$$5x^2 - 3x + 4 - 2x^2 + 7$$

$$3x^2 - 3x + 11$$

31. D: SOHCAHTOA is used to find the missing side length. Because the angle and adjacent side are known, $\tan 60 = \frac{x}{13}$. Making sure to evaluate tangent with an argument in degrees, this equation gives $x = 13 \tan 60 = 22.52$.

32. C: The sample space is made up of $8 + 7 + 6 + 5 = 26$ balls. The probability of pulling each individual ball is $\frac{1}{26}$. Since there are 7 yellow balls, the probability of pulling a yellow ball is $\frac{7}{26}$.

33. D: $x \leq -5$. When solving a linear equation or inequality:

Distribution is performed if necessary:

$$-3(x + 4) \rightarrow -3x - 12 \geq x + 8$$

This means that any like terms on the same side of the equation/inequality are combined.

The equation/inequality is manipulated to get the variable on one side. In this case, subtracting x from both sides produces:

$$-4x - 12 \geq 8$$

The variable is isolated using inverse operations to undo addition/subtraction. Adding 12 to both sides produces $-4x \geq 20$.

The variable is isolated using inverse operations to undo multiplication/division. Remember if dividing by a negative number, the relationship of the inequality reverses, so the sign is flipped. In this case, dividing by -4 on both sides produces $x \leq -5$.

34. A: If each man gains 10 pounds, every original data point will increase by 10 pounds. Therefore, the man with the original median will still have the median value, but that value will increase by 10. The smallest value and largest value will also increase by 10 and, therefore, the difference between the two won't change. The range does not change in value and, thus, remains the same.

35. D: When an ordered pair is reflected over an axis, the sign of one of the coordinates must change. When it's reflected over the x-axis, the sign of the y coordinate must change. The x value remains the same. Therefore, the new ordered pair is $(-3, 4)$.

36. A: Because the volume of the given sphere is 288π cubic meters, this gives:

$$^4/_3 \pi r^3 = 288\pi$$

This equation is solved for r to obtain a radius of 6 meters. The formula for surface area is $4\pi r^2$ so:

$$SA = 4\pi 6^2 = 144\pi \text{ square meters}$$

37. B: A rectangle is a specific type of parallelogram. It has 4 right angles. A square is a rhombus that has 4 right angles. Therefore, a square is always a rectangle because it has two sets of parallel lines and 4 right angles.

38. B: The sine of 30° is equal to ½. Choice *A* is not the correct answer because the sine of 15° is .2588. Choice *C* is not the answer because the sine of 45° is .707. Choice *D* is not the answer because the sine of 90 degrees is 1.

39. C: $Chord\ Length = 2 \times radius \times \sin\frac{angle}{2}$, and $2 \times 30 \times \sin 30 = 30$. Choice *A* is not the correct answer because that is the radius divided by 6. Choice *B* is not the correct answer because that is the radius divided by 3. Choice *D* is not the correct answer because that is the radius minus 10.

40. C: Graphing the function $y = \cos(x)$ shows that the curve starts at $(0, 1)$, has an amplitude of 2, and a period of 2π. This same curve can be constructed using the sine graph, by shifting the graph to the left $\frac{\pi}{2}$ units. This equation is in the form $y = \sin(x + \frac{\pi}{2})$.

41. A: Every 8 ml of medicine requires 5 mL. The 45 mL first needs to be split into portions of 8 mL. This results in $\frac{45}{8}$ portions. Each portion requires 5 mL. Therefore, $\frac{45}{8} \times 5 = \frac{45*5}{8} = \frac{225}{8}$ mL is necessary.

42. D: The midpoint formula should be used.

$$M = \left(\frac{x_1 + x_2}{2}, \frac{y_1 + y_2}{2}\right) = \left(\frac{-1 + 3}{2}, \frac{2 + (-6)}{2}\right) = (1, -2)$$

43. A: First, the sample mean must be calculated.

$$\bar{x} = \frac{1}{4}(1 + 3 + 5 + 7) = 4$$

The standard deviation of the data set is $\sigma = \sqrt{\frac{\sum(x-\bar{x})^2}{n-1}}$, and $n = 4$ represents the number of data points.

Therefore:

$$\sigma = \sqrt{\frac{1}{3}[(1-4)^2 + (3-4)^2 + (5-4)^2 + (7-4)^2]} = \sqrt{\frac{1}{3}(9+1+1+9)} = 2.58$$

44. B: An equilateral triangle has three sides of equal length, so if the total perimeter is 18 feet, each side must be 6 feet long. A square with sides of 6 feet will have an area of $6^2 = 36$ square feet.

45. B: 300 miles in 4 hours is $\frac{300}{4}$ = 75 miles per hour. In 1.5 hours, the car will go 1.5×75 miles, or 112.5 miles.

46. A: Let the unknown score be x. The average will be:

$$\frac{5 \times 50 + 4 \times 70 + x}{10} = \frac{530 + x}{10} = 55$$

Multiply both sides by 10 to get $530 + x = 550$, or $x = 20$.

47. A: The formula for the area of the circle is πr^2, and 9 squared is 81. Choice *B* is not the correct answer because that is 2×9. Choice *C* is not the correct answer because that is 9×10. Choice *D* is not the correct answer because that is simply the value of the radius.

48. B: 13,078. The power of 10 by which a digit is multiplied corresponds with the number of zeros following the digit when expressing its value in standard form. Therefore:

$$(1 \times 10^4) + (3 \times 10^3) + (7 \times 10^1) + (8 \times 10^0)$$

$$10,000 + 3,000 + 70 + 8 = 13,078$$

49. A: Using the trigonometric identity $\tan(\theta) = \frac{\sin(\theta)}{\cos(\theta)}$, the expression becomes $\frac{\sin\theta}{\cos\theta}\cos\theta$. The factors that are the same on the top and bottom cancel out, leaving the simplified expression $\sin\theta$.

50. B: $\frac{11}{15}$. Fractions must have like denominators to be added. The least common multiple of the denominators 3 and 5 is found. The LCM is 15, so both fractions should be changed to equivalent fractions with a denominator of 15. To determine the numerator of the new fraction, the old numerator is multiplied by the same number by which the old denominator is multiplied to obtain the new denominator. For the fraction $\frac{1}{3}$, 3 multiplied by 5 will produce 15. Therefore, the numerator is multiplied by 5 to produce the new numerator $\left(\frac{1\times5}{3\times5} = \frac{5}{15}\right)$. For the fraction $\frac{2}{5}$, multiplying both the

numerator and denominator by 3 produces $\frac{6}{15}$. When fractions have like denominators, they are added by adding the numerators and keeping the denominator the same:

$$\frac{5}{15} + \frac{6}{15} = \frac{11}{15}$$

51. B: $90° - 30° = 60°$. Choice *A* is not the correct answer because that is simply the original angle given. Choice *C* is not the correct answer since that is the angle you subtract from. Choice *D* is not the correct answer because that is $90° + 30°$.

52. A: $90° - 60° = 30°$. Choice *B* is not the correct answer because this is simply the original angle given. Choice *C* is not the correct answer since that is the angle you subtract from. Choice *D* is not the correct answer because that is $90° + 30°$.

53.

6	.	2	5

If a line is parallel to a side of a triangle and intersects the other two sides of the triangle, it separates the sides into corresponding segments of proportional lengths. To solve, set up a proportion:

$$\frac{AE}{AD} = \frac{AB}{AC} \rightarrow \frac{4}{5} = \frac{5}{x}$$

Cross multiplying yields:

$$4x = 25 \rightarrow x = 6.25$$

54.

	0	.	1	2
⊝	⊝	⊝	⊝	⊝
	⊙	●	⊙	⊙
	●	0	0	0
	1	1	●	1
	2	2	2	●
	3	3	3	3
	4	4	4	4
	5	5	5	5
	6	6	6	6
	7	7	7	7
	8	8	8	8
	9	9	9	9

The fraction is converted so that the denominator is 100 by multiplying the numerator and denominator by 4, to get $\frac{3}{25} = \frac{12}{100}$. Dividing a number by 100 just moves the decimal point two places to the left, with a result of 0.12.

55.

			2	0
⊝	⊝	⊝	⊝	⊝
	⊙	⊙	⊙	⊙
	0	0	0	●
	1	1	1	1
	2	2	●	2
	3	3	3	3
	4	4	4	4
	5	5	5	5
	6	6	6	6
	7	7	7	7
	8	8	8	8
	9	9	9	9

30% is $\frac{3}{10}$. The number itself must be $\frac{10}{3}$ of 6, or $\frac{10}{3} \times 6 = 10 \times 2 = 20$.

56.

			1	0
-	-	-	-	-
	.	.	.	.
	0	0	0	●
	1	1	●	1
	2	2	2	2
	3	3	3	3
	4	4	4	4
	5	5	5	5
	6	6	6	6
	7	7	7	7
	8	8	8	8
	9	9	9	9

8 squared is 64, and 6 squared is 36. These should be added together to get $64 + 36 = 100$. Then, the last step is to find the square root of 100 which is 10.

PSAT Practice Test #2

Reading

U.S. Literature Passage #1

Objective: Read and comprehend an excerpt from U.S. Literature

Questions 1-6 are based upon the following passage:

This excerpt is an adaptation of Jonathan Swift's *Gulliver's Travels into Several Remote Nations of the World.*

My gentleness and good behaviour had gained so far on the emperor and his court, and indeed upon the army and people in general, that I began to conceive hopes of getting my liberty in a short time. I took all possible methods to cultivate this favourable disposition. The natives came, by degrees, to be less apprehensive of any danger from me. I would sometimes lie down, and let five or six of them dance on my hand; and at last the boys and girls would venture to come and play at hide-and-seek in my hair. I had now made a good progress in understanding and speaking the language. The emperor had a mind one day to entertain me with several of the country shows, wherein they exceed all nations I have known, both for dexterity and magnificence. I was diverted with none so much as that of the rope-dancers, performed upon a slender white thread, extended about two feet, and twelve inches from the ground. Upon which I shall desire liberty, with the reader's patience, to enlarge a little.

This diversion is only practised by those persons who are candidates for great employments, and high favour at court. They are trained in this art from their youth, and are not always of noble birth, or liberal education. When a great office is vacant, either by death or disgrace (which often happens,) five or six of those candidates petition the emperor to entertain his majesty and the court with a dance on the rope; and whoever jumps the highest, without falling, succeeds in the office. Very often the chief ministers themselves are commanded to show their skill, and to convince the emperor that they have not lost their faculty. Flimnap, the treasurer, is allowed to cut a caper on the straight rope, at least an inch higher than any other lord in the whole empire. I have seen him do the summerset several times together, upon a trencher fixed on a rope which is no thicker than a common packthread in England. My friend Reldresal, principal secretary for private affairs, is, in my opinion, if I am not partial, the second after the treasurer; the rest of the great officers are much upon a par.

1. Which of the following statements best summarize the central purpose of this text?
   a. Gulliver details his fondness for the archaic yet interesting practices of his captors.
   b. Gulliver conjectures about the intentions of the aristocratic sector of society.
   c. Gulliver becomes acquainted with the people and practices of his new surroundings.
   d. Gulliver's differences cause him to become penitent around new acquaintances.

2. What is the word *principal* referring to in the following text?

> My friend Reldresal, principal secretary for private affairs, is, in my opinion, if I am not partial, the second after the treasurer; the rest of the great officers are much upon a par.

    a. Primary or chief
    b. An acolyte
    c. An individual who provides nurturing
    d. One in a subordinate position

3. What can the reader infer from this passage?

> I would sometimes lie down, and let five or six of them dance on my hand; and at last the boys and girls would venture to come and play at hide-and-seek in my hair.

    a. The children tortured Gulliver.
    b. Gulliver traveled because he wanted to meet new people.
    c. Gulliver is considerably larger than the children who are playing around him.
    d. Gulliver has a genuine love and enthusiasm for people of all sizes.

4. What is the significance of the word *mind* in the following passage?

> The emperor had a mind one day to entertain me with several of the country shows, wherein they exceed all nations I have known, both for dexterity and magnificence.

    a. The ability to think
    b. A collective vote
    c. A definitive decision
    d. A mythological question

5. Which of the following assertions does NOT support the fact that games are a commonplace event in this culture?
    a. My gentlest and good behavior . . . short time.
    b. They are trained in this art from their youth . . . liberal education.
    c. Very often the chief ministers themselves are commanded to show their skill . . . not lost their faculty.
    d. Flimnap, the treasurer, is allowed to cut a caper on the straight rope . . . higher than any other lord in the whole empire.

6. How do the roles of Flimnap and Reldresal serve as evidence of the community's emphasis in regards to the correlation between physical strength and leadership abilities?
    a. Only children used Gulliver's hands as a playground.
    b. The two men who exhibited superior abilities held prominent positions in the community.
    c. Only common townspeople, not leaders, walk the straight rope.
    d. No one could jump higher than Gulliver.

Literature Passage #2

Objective: Read and comprehend an excerpt from U.S. Literature.

Questions 7-12 are based upon the following passage:

This excerpt is an adaptation of Robert Louis Stevenson's *The Strange Case of Dr. Jekyll and Mr. Hyde.*

"Did you ever come across a protégé of his—one Hyde?" He asked.

"Hyde?" repeated Lanyon. "No. Never heard of him. Since my time."

That was the amount of information that the lawyer carried back with him to the great, dark bed on which he tossed to and fro until the small hours of the morning began to grow large. It was a night of little ease to his toiling mind, toiling in mere darkness and besieged by questions.

Six o'clock struck on the bells of the church that was so conveniently near to Mr. Utterson's dwelling, and still he was digging at the problem. Hitherto it had touched him on the intellectual side alone; but; but now his imagination also was engaged, or rather enslaved; and as he lay and tossed in the gross darkness of the night in the curtained room, Mr. Enfield's tale went by before his mind in a scroll of lighted pictures. He would be aware of the great field of lamps in a nocturnal city; then of the figure of a man walking swiftly; then of a child running from the doctor's; and then these met, and that human Juggernaut trod the child down and passed on regardless of her screams. Or else he would see a room in a rich house, where his friend lay asleep, dreaming and smiling at his dreams; and then the door of that room would be opened, the curtains of the bed plucked apart, the sleeper recalled, and, lo! There would stand by his side a figure to whom power was given, and even at that dead hour he must rise and do its bidding. The figure in these two phrases haunted the lawyer all night; and if at anytime he dozed over, it was but to see it glide more stealthily through sleeping houses, or move the more swiftly, and still the more smoothly, even to dizziness, through wider labyrinths of lamplighted city, and at every street corner crush a child and leave her screaming. And still the figure had no face by which he might know it; even in his dreams it had no face, or one that baffled him and melted before his eyes; and thus there it was that there sprung up and grew apace in the lawyer's mind a singularly strong, almost an inordinate, curiosity to behold the features of the real Mr. Hyde. If he could but once set eyes on him, he thought the mystery would lighten and perhaps roll altogether away, as was the habit of mysterious things when well examined. He might see a reason for his friend's strange preference or bondage, and even for the startling clauses of the will. And at least it would be a face worth seeing: the face of a man who was without bowels of mercy: a face which had but to show itself to raise up, in the mind of the unimpressionable Enfield, a spirit of enduring hatred.

From that time forward, Mr. Utterson began to haunt the door in the by street of shops. In the morning before office hours, at noon when business was plenty of time scarce, at night under the face of the full city moon, by all lights and at all hours of solitude or concourse, the lawyer was to be found on his chosen post.

"If he be Mr. Hyde," he had thought, "I should be Mr. Seek."

7. What is the purpose of the use of repetition in the following passage?

It was a night of little ease to his toiling mind, toiling in mere darkness and besieged by questions.

a. It serves as a demonstration of the mental state of Mr. Lanyon.
b. It is reminiscent of the church bells that are mentioned in the story.
c. It mimics Mr. Utterson's ambivalence.
d. It emphasizes Mr. Utterson's anguish in failing to identify Hyde's whereabouts.

8. What is the setting of the story in this passage?
a. In the city
b. On the countryside
c. In a jail
d. In a mental health facility

9. What can one infer about the meaning of the word "Juggernaut" from the author's use of it in the passage?
a. It is an apparition that appears at daybreak.
b. It scares children.
c. It is associated with space travel.
d. Mr. Utterson finds it soothing.

10. What is the definition of the word *haunt* in the following passage?

From that time forward, Mr. Utterson began to haunt the door in the by street of shops. In the morning before office hours, at noon when business was plenty of time scarce, at night under the face of the full city moon, by all lights and at all hours of solitude or concourse, the lawyer was to be found on his chosen post.

a. To levitate
b. To constantly visit
c. To terrorize
d. To daunt

11. The phrase *labyrinths of lamplighted city* contains an example of what?
a. Hyperbole
b. Simile
c. Juxtaposition
d. Alliteration

12. What can one reasonably conclude from the final comment of this passage?

"If he be Mr. Hyde," he had thought, "I should be Mr. Seek."

a. The speaker is considering a name change.
b. The speaker is experiencing an identity crisis.
c. The speaker has mistakenly been looking for the wrong person.
d. The speaker intends to continue to look for Hyde.

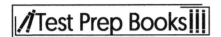

History Passage #1

Objective: Read and comprehend an excerpt written from a historical perspective.

Questions 13-18 are based upon the following passage:

This excerpt is adapted from "What to the Slave is the Fourth of July?" Rochester, New York July 5, 1852

Fellow citizens—Pardon me, and allow me to ask, why am I called upon to speak here today? What have I, or those I represent, to do with your national independence? Are the great principles of political freedom and of natural justice, embodied in that Declaration of Independence, extended to us? And am I therefore called upon to bring our humble offering to the national altar, and to confess the benefits, and express devout gratitude for the blessings, resulting from your independence to us?

Would to God, both for your sakes and ours, ours that an affirmative answer could be truthfully returned to these questions! Then would my task be light, and my burden easy and delightful. For who is there so cold that a nation's sympathy could not warm him? Who so obdurate and dead to the claims of gratitude that would not thankfully acknowledge such priceless benefits? Who so stolid and selfish, that would not give his voice to swell the hallelujahs of a nation's jubilee, when the chains of servitude had been torn from his limbs? I am not that man. In a case like that, the dumb may eloquently speak, and the lame man leap as an hart.

But, such is not the state of the case. I say it with a sad sense of the disparity between us. I am not included within the pale of this glorious anniversary. Oh pity! Your high independence only reveals the immeasurable distance between us. The blessings in which you this day rejoice, I do not enjoy in common. The rich inheritance of justice, liberty, prosperity, and independence, bequeathed by your fathers, is shared by *you*, not by *me*. This Fourth of July is *yours,* not *mine*. You may rejoice, *I* must mourn. To drag a man in fetters into the grand illuminated temple of liberty, and call upon him to join you in joyous anthems, were inhuman mockery and sacrilegious irony. Do you mean, citizens, to mock me, by asking me to speak today? If so there is a parallel to your conduct. And let me warn you that it is dangerous to copy the example of a nation whose crimes, towering up to heaven, were thrown down by the breath of the Almighty, burying that nation and irrecoverable ruin! I can today take up the plaintive lament of a peeled and woe-smitten people.

By the rivers of Babylon, there we sat down. Yea! We wept when we remembered Zion. We hanged our harps upon the willows in the midst thereof. For there, they that carried us away captive, required of us a song; and they who wasted us required of us mirth, saying, "Sing us one of the songs of Zion." How can we sing the Lord's song in a strange land? If I forget thee, O Jerusalem, let my right hand forget her cunning. If I do not remember thee, let my tongue cleave to the roof of my mouth.

13. What is the tone of the first paragraph of this passage?
    a. Exasperated
    b. Inclusive
    c. Contemplative
    d. Nonchalant

14. Which word CANNOT be used synonymously with the term *obdurate* as it is conveyed in the text below?

> Who so obdurate and dead to the claims of gratitude, that would not thankfully acknowledge such priceless benefits?

a. Steadfast
b. Stubborn
c. Contented
d. Unwavering

15. What is the central purpose of this text?
a. To demonstrate the author's extensive knowledge of the Bible
b. To address the hypocrisy of the Fourth of July holiday
c. To convince wealthy landowners to adopt new holiday rituals
d. To explain why minorities often relished the notion of segregation in government institutions

16. Which statement serves as evidence of the question above?
a. By the rivers of Babylon . . . down.
b. Fellow citizens . . . today.
c. I can . . . woe-smitten people.
d. The rich inheritance of justice . . . *not by me.*

17. The statement below features an example of which of the following literary devices?

> Oh pity! Your high independence only reveals the immeasurable distance between us.

a. Assonance
b. Parallelism
c. Amplification
d. Hyperbole

18. The speaker's use of biblical references, such as "rivers of Babylon" and the "songs of Zion," helps the reader to do all of the following EXCEPT:
a. Identify with the speaker through the use of common text.
b. Convince the audience that injustices have been committed by referencing another group of people who have been previously affected by slavery.
c. Display the equivocation of the speaker and those that he represents.
d. Appeal to the listener's sense of humanity.

History Passage #2

Objective: Read and comprehend an excerpt written from a historical perspective.

Questions 19-24 are based upon the following passage:

This excerpt is an adaptation from Abraham Lincoln's Address Delivered at the Dedication of the Cemetery at Gettysburg, November 19, 1863.

> Four score and seven years ago our fathers brought forth on this continent, a new nation, conceived in liberty, and dedicated to the proposition that all men are created equal.

Now we are engaged in a great civil war, testing whether that nation, or any nation so conceived and so dedicated, can long endure. We are met on a great battlefield of that war. We have come to dedicate a portion of that field, as a final resting place for those who here gave their lives that this nation might live. It is altogether fitting and proper that we should do this.

But, in a larger sense, we cannot dedicate—we cannot consecrate that we cannot hallow—this ground. The brave men, living and dead, who struggled here, have consecrated it, far above our poor power to add or detract. The world will little note, nor long remember what we say here, but it can never forget what they did here. It is for us the living, rather, to be dedicated here to the unfinished work which they who fought here have thus far so nobly advanced. It is rather for us to be here and dedicated to the great task remaining before us—that from these honored dead we take increased devotion to that cause for which they gave the last full measure of devotion—that we here highly resolve that these dead shall not have died in vain—that these this nation, under God, shall have a new birth of freedom—and that government of people, by the people, for the people, shall not perish from the earth.

19. The best description for the phrase *four score and seven years ago* is which of the following?
    a. A unit of measurement
    b. A period of time
    c. A literary movement
    d. A statement of political reform

20. What is the setting of this text?
    a. A battleship off of the coast of France
    b. A desert plain on the Sahara Desert
    c. A battlefield in North America
    d. The residence of Abraham Lincoln

21. Which war is Abraham Lincoln referring to in the following passage?
    Now we are engaged in a great civil war, testing whether that nation, or any nation so conceived and so dedicated, can long endure.

    a. World War I
    b. The War of the Spanish Succession
    c. World War II
    d. The American Civil War

22. What message is the author trying to convey through this address?
    a. The audience should perpetuate the ideals of freedom that the soldiers died fighting for.
    b. The audience should honor the dead by establishing an annual memorial service.
    c. The audience should form a militia that would overturn the current political structure.
    d. The audience should forget the lives that were lost and discredit the soldiers.

23. Which rhetorical device is being used in the following passage?
    . . . we here highly resolve that these dead shall not have died in vain—that these this nation, under God, shall have a new birth of freedom—and that government of people, by the people, for the people, shall not perish from the earth.

a. Antimetabole
b. Antiphrasis
c. Anaphora
d. Epiphora

24. What is the effect of Lincoln's statement in the following passage?

> But, in a larger sense, we cannot dedicate—we cannot consecrate that we cannot hallow—this ground. The brave men, living and dead, who struggled here, have consecrated it, far above our poor power to add or detract.

a. His comparison emphasizes the great sacrifice of the soldiers who fought in the war.
b. His comparison serves as a reminder of the inadequacies of his audience.
c. His comparison serves as a catalyst for guilt and shame among audience members.
d. His comparison attempts to illuminate the great differences between soldiers and civilians.

History Passage #3

Objective: Read and comprehend an excerpt written from a historical perspective.

Questions 25-30 are based upon the following passage:

This excerpt is adapted from Charles Dickens' speech in Birmingham in England on December 30, 1853 on behalf of the Birmingham and Midland Institute.

> My Good Friends,—When I first imparted to the committee of the projected Institute my particular wish that on one of the evenings of my readings here the main body of my audience should be composed of working men and their families, I was animated by two desires; first, by the wish to have the great pleasure of meeting you face to face at this Christmas time, and accompany you myself through one of my little Christmas books; and second, by the wish to have an opportunity of stating publicly in your presence, and in the presence of the committee, my earnest hope that the Institute will, from the beginning, recognise one great principle—strong in reason and justice—which I believe to be essential to the very life of such an Institution. It is, that the working man shall, from the first unto the last, have a share in the management of an Institution which is designed for his benefit, and which calls itself by his name.

> I have no fear here of being misunderstood—of being supposed to mean too much in this. If there ever was a time when any one class could of itself do much for its own good, and for the welfare of society—which I greatly doubt—that time is unquestionably past. It is in the fusion of different classes, without confusion; in the bringing together of employers and employed; in the creating of a better common understanding among those whose interests are identical, who depend upon each other, who are vitally essential to each other, and who never can be in unnatural antagonism without deplorable results, that one of the chief principles of a Mechanics' Institution should consist. In this world a great deal of the bitterness among us arises from an imperfect understanding of one another. Erect in Birmingham a great Educational Institution, properly educational; educational of the feelings as well as of the reason; to which all orders of Birmingham men contribute; in which all orders of Birmingham men meet; wherein all orders of Birmingham men are faithfully

represented—and you will erect a Temple of Concord here which will be a model edifice to the whole of England.

Contemplating as I do the existence of the Artisans' Committee, which not long ago considered the establishment of the Institute so sensibly, and supported it so heartily, I earnestly entreat the gentlemen—earnest I know in the good work, and who are now among us,—by all means to avoid the great shortcoming of similar institutions; and in asking the working man for his confidence, to set him the great example and give him theirs in return. You will judge for yourselves if I promise too much for the working man, when I say that he will stand by such an enterprise with the utmost of his patience, his perseverance, sense, and support; that I am sure he will need no charitable aid or condescending patronage; but will readily and cheerfully pay for the advantages which it confers; that he will prepare himself in individual cases where he feels that the adverse circumstances around him have rendered it necessary; in a word, that he will feel his responsibility like an honest man, and will most honestly and manfully discharge it. I now proceed to the pleasant task to which I assure you I have looked forward for a long time.

25. Which word is most closely synonymous with the word *patronage* as it appears in the following statement?

. . . that I am sure he will need no charitable aid or condescending patronage

a. Auspices
b. Aberration
c. Acerbic
d. Adulation

26. Which term is most closely aligned with the definition of the term *working man* as it is defined in the following passage?

You will judge for yourselves if I promise too much for the working man, when I say that he will stand by such an enterprise with the utmost of his patience, his perseverance, sense, and support . . .

a. Plebeian
b. Viscount
c. Entrepreneur
d. Bourgeois

27. Which of the following statements most closely correlates with the definition of the term *working man* as it is defined in Question 26?

a. A working man is not someone who works for institutions or corporations, but someone who is well versed in the workings of the soul.
b. A working man is someone who is probably not involved in social activities because the physical demand for work is too high.
c. A working man is someone who works for wages among the middle class.
d. The working man has historically taken to the field, to the factory, and now to the screen.

28. Based upon the contextual evidence provided in the passage above, what is the meaning of the term *enterprise* in the third paragraph?
    a. Company
    b. Courage
    c. Game
    d. Cause

29. The speaker addresses his audience as *My Good Friends*—what kind of credibility does this salutation give to the speaker?
    a. The speaker is an employer addressing his employees, so the salutation is a way for the boss to bridge the gap between himself and his employees.
    b. The speaker's salutation is one from an entertainer to his audience and uses the friendly language to connect to his audience before a serious speech.
    c. The salutation gives the serious speech that follows a somber tone, as it is used ironically.
    d. The speech is one from a politician to the public, so the salutation is used to grab the audience's attention.

30. According to the aforementioned passage, what is the speaker's second desire for his time in front of the audience?
    a. To read a Christmas story
    b. For the working man to have a say in his institution which is designed for his benefit
    c. To have an opportunity to stand in their presence
    d. For the life of the institution to be essential to the audience as a whole

Science Passage #1

Objective: Read and comprehend an excerpt written from a scientific perspective.

Questions 31-36 are based upon the following passage:

This excerpt is adapted from *Our Vanishing Wildlife,* by William T. Hornaday

> Three years ago, I think there were not many bird-lovers in the United States, who believed it possible to prevent the total extinction of both egrets from our fauna. All the known rookeries accessible to plume-hunters had been totally destroyed. Two years ago, the secret discovery of several small, hidden colonies prompted William Dutcher, President of the National Association of Audubon Societies, and Mr. T. Gilbert Pearson, Secretary, to attempt the protection of those colonies. With a fund contributed for the purpose, wardens were hired and duly commissioned. As previously stated, one of those wardens was shot dead in cold blood by a plume hunter. The task of guarding swamp rookeries from the attacks of money-hungry desperadoes to whom the accursed plumes were worth their weight in gold, is a very chancy proceeding. There is now one warden in Florida who says that "before they get my rookery they will first have to get me."
>
> Thus far the protective work of the Audubon Association has been successful. Now there are twenty colonies, which contain all told, about 5,000 egrets and about 120,000 herons and ibises which are guarded by the Audubon wardens. One of the most important is on Bird Island, a mile out in Orange Lake, central Florida, and it is ably defended by Oscar E. Baynard. To-day, the plume hunters who do not dare to raid the guarded rookeries are trying to study out the lines of flight of the birds, to and from

their feeding-grounds, and shoot them in transit. Their motto is—"Anything to beat the law, and get the plumes." It is there that the state of Florida should take part in the war.

The success of this campaign is attested by the fact that last year a number of egrets were seen in eastern Massachusetts—for the first time in many years. And so to-day the question is, can the wardens continue to hold the plume-hunters at bay?

31. The author's use of first person pronoun in the following text does NOT have which of the following effects?

Three years ago, I think there were not many bird-lovers in the United States, who believed it possible to prevent the total extinction of both egrets from our fauna.

a. The phrase *I think* acts as a sort of hedging, where the author's tone is less direct and/or absolute.
b. It allows the reader to more easily connect with the author.
c. It encourages the reader to empathize with the egrets.
d. It distances the reader from the text by overemphasizing the story.

32. What purpose does the quote serve at the end of the first paragraph?
a. The quote shows proof of a hunter threatening one of the wardens.
b. The quote lightens the mood by illustrating the colloquial language of the region.
c. The quote provides an example of a warden protecting one of the colonies.
d. The quote provides much needed comic relief in the form of a joke.

33. What is the meaning of the word *rookeries* in the following text?

To-day, the plume hunters who do not dare to raid the guarded rookeries are trying to study out the lines of flight of the birds, to and from their feeding-grounds, and shoot them in transit.

a. Houses in a slum area
b. A place where hunters gather to trade tools
c. A place where wardens go to trade stories
d. A colony of breeding birds

34. What is on Bird Island?
a. Hunters selling plumes
b. An important bird colony
c. Bird Island Battle between the hunters and the wardens
d. An important egret with unique plumes

35. What is the main purpose of the passage?
a. To persuade the audience to act in preservation of the bird colonies
b. To show the effect hunting egrets has had on the environment
c. To argue that the preservation of bird colonies has had a negative impact on the environment
d. To demonstrate the success of the protective work of the Audubon Association

36. Why are hunters trying to study the lines of flight of the birds?
a. To study ornithology, one must know the lines of flight that birds take.
b. To help wardens preserve the lives of the birds
c. To have a better opportunity to hunt the birds
d. To build their homes under the lines of flight because they believe it brings good luck

Science Passage #2

Objective: Read and comprehend an excerpt written from a scientific perspective.

Questions 37-42 are based upon the following passage:

This excerpt is adapted from *The Life-Story of Insects,* by Geo H. Carpenter.

Insects as a whole are preeminently creatures of the land and the air. This is shown not only by the possession of wings by a vast majority of the class, but by the mode of breathing to which reference has already been made, a system of branching air-tubes carrying atmospheric air with its combustion-supporting oxygen to all the insect's tissues. The air gains access to these tubes through a number of paired air-holes or spiracles, arranged segmentally in series.

It is of great interest to find that, nevertheless, a number of insects spend much of their time under water. This is true of not a few in the perfect winged state, as for example aquatic beetles and water-bugs ('boatmen' and 'scorpions') which have some way of protecting their spiracles when submerged, and, possessing usually the power of flight, can pass on occasion from pond or stream to upper air. But it is advisable in connection with our present subject to dwell especially on some insects that remain continually under water till they are ready to undergo their final moult and attain the winged state, which they pass entirely in the air. The preparatory instars of such insects are aquatic; the adult instar is aerial. All may-flies, dragon-flies, and caddis-flies, many beetles and two-winged flies, and a few moths thus divide their life-story between the water and the air. For the present we confine attention to the Stone-flies, the May-flies, and the Dragon-flies, three well-known orders of insects respectively called by systematists the Plecoptera, the Ephemeroptera and the Odonata.

In the case of many insects that have aquatic larvae, the latter are provided with some arrangement for enabling them to reach atmospheric air through the surface-film of the water. But the larva of a stone-fly, a dragon-fly, or a may-fly is adapted more completely than these for aquatic life; it can, by means of gills of some kind, breathe the air dissolved in water.

37. Which statement best details the central idea in this passage?
    a. It introduces certain insects that transition from water to air.
    b. It delves into entomology, especially where gills are concerned.
    c. It defines what constitutes as insects' breathing.
    d. It invites readers to have a hand in the preservation of insects.

38. Which definition most closely relates to the usage of the word *moult* in the passage?
    a. An adventure of sorts, especially underwater
    b. Mating act between two insects
    c. The act of shedding part or all of the outer shell
    d. Death of an organism that ends in a revival of life

39. What is the purpose of the first paragraph in relation to the second paragraph?
    a. The first paragraph serves as a cause, and the second paragraph serves as an effect.
    b. The first paragraph serves as a contrast to the second.
    c. The first paragraph is a description for the argument in the second paragraph.
    d. The first and second paragraphs are merely presented in a sequence.

40. What does the following sentence most nearly mean?
    The preparatory instars of such insects are aquatic; the adult instar is aerial.

    a. The volume of water is necessary to prep the insect for transition rather than the volume of the air.
    b. The abdomen of the insect is designed like a star in the water as well as the air.
    c. The stage of preparation in between molting is acted out in the water, while the last stage is in the air.
    d. These insects breathe first in the water through gills, yet continue to use the same organs to breathe in the air.

41. Which of the statements reflect information that one could reasonably infer based on the author's tone?
    a. The author's tone is persuasive and attempts to call the audience to action.
    b. The author's tone is passionate due to excitement over the subject and personal narrative.
    c. The author's tone is informative and exhibits interest in the subject of the study.
    d. The author's tone is somber, depicting some anger at the state of insect larvae.

42. Which statement best describes stoneflies, mayflies, and dragonflies?
    a. They are creatures of the land and the air.
    b. They have a way of protecting their spiracles when submerged.
    c. Their larvae can breathe the air dissolved in water through gills of some kind.
    d. The preparatory instars of these insects are aerial.

Science Passage #3

Objective: Read and comprehend an excerpt written from a scientific perspective.

Questions 43-47 are based upon the following passage:

This excerpt is adapted from "The 'Hatchery' of the Sun-Fish"--- *Scientific American,* #711

> I have thought that an example of the intelligence (instinct?) of a class of fish which has come under my observation during my excursions into the Adirondack region of New York State might possibly be of interest to your readers, especially as I am not aware that any one except myself has noticed it, or, at least, has given it publicity.
>
> The female sun-fish (called, I believe, in England, the roach or bream) makes a "hatchery" for her eggs in this wise. Selecting a spot near the banks of the numerous lakes in which this region abounds, and where the water is about 4 inches deep, and still, she builds, with her tail and snout, a circular embankment 3 inches in height and 2 thick. The circle, which is as perfect a one as could be formed with mathematical instruments, is usually a foot and a half in diameter; and at one side of this circular wall an opening is left by the fish of just sufficient width to admit her body.

The mother sun-fish, having now built or provided her "hatchery," deposits her spawn within the circular inclosure, and mounts guard at the entrance until the fry are hatched out and are sufficiently large to take charge of themselves. As the embankment, moreover, is built up to the surface of the water, no enemy can very easily obtain an entrance within the inclosure from the top; while there being only one entrance, the fish is able, with comparative ease, to keep out all intruders.

I have, as I say, noticed this beautiful instinct of the sun-fish for the perpetuity of her species more particularly in the lakes of this region; but doubtless the same habit is common to these fish in other waters.

43. What is the purpose of this passage?
    a. To show the effects of fish hatcheries on the Adirondack region
    b. To persuade the audience to study Ichthyology (fish science)
    c. To depict the sequence of mating among sun-fish
    d. To enlighten the audience on the habits of sun-fish and their hatcheries

44. What does the word *wise* in this passage most closely mean?
    a. Knowledge
    b. Manner
    c. Shrewd
    d. Ignorance

45. What is the definition of the word *fry* as it appears in the following passage?
    The mother sun-fish, having now built or provided her "hatchery," deposits her spawn within the circular inclosure, and mounts guard at the entrance until the fry are hatched out and are sufficiently large to take charge of themselves.

    a. Fish at the stage of development where they are capable of feeding themselves
    b. Fish eggs that have been fertilized
    c. A place where larvae is kept out of danger from other predators
    d. A dish where fish is placed in oil and fried until golden brown

46. How is the circle that keeps the larvae of the sun-fish made?
    a. It is formed with mathematical instruments.
    b. The sun-fish builds it with her tail and snout.
    c. It is provided to her as a "hatchery" by Mother Nature.
    d. The sun-fish builds it with her larvae.

47. The author included the third paragraph in the following passage to achieve which of the following effects?
    a. To complicate the subject matter
    b. To express a bias
    c. To insert a counterargument
    d. To conclude a sequence and add a final detail

Writing

Read the selection about travelling in an RV and answer Questions 1-7.

I have to admit that when my father bought a recreational vehicle (RV), I thought he was making a huge mistake. I didn't really know anything about RVs, but I knew that my dad was as big a "city slicker" as there was. (1) <u>In fact, I even thought he might have gone a little bit crazy.</u> On trips to the beach, he preferred to swim at the pool, and whenever he went hiking, he avoided touching any plants for fear that they might be poison ivy. Why would this man, with an almost irrational fear of the outdoors, want a 40-foot camping behemoth?

(2) <u>The RV</u> was a great purchase for our family and brought us all closer together. Every morning (3) <u>we would wake up, eat breakfast, and broke camp.</u> We laughed at our own comical attempts to back The Beast into spaces that seemed impossibly small. (4) <u>We rejoiced as "hackers."</u> When things inevitably went wrong and we couldn't solve the problems on our own, we discovered the incredible helpfulness and friendliness of the RV community. (5) <u>We even made some new friends in the process.</u>

(6) <u>Above all, it allowed us to share adventures. While travelling across America</u>, which we could not have experienced in cars and hotels. Enjoying a campfire on a chilly summer evening with the mountains of Glacier National Park in the background, or waking up early in the morning to see the sun rising over the distant spires of Arches National Park are memories that will always stay with me and our entire family. (7) <u>Those are also memories that my siblings and me</u> have now shared with our own children.

1. Which of the following would be the best choice for this sentence (reproduced below)?

   In fact, I even thought he might have gone a little bit crazy.

   a. NO CHANGE
   b. Move the sentence so that it comes before the preceding sentence.
   c. Move the sentence to the end of the first paragraph.
   d. Omit the sentence.

2. Which of the following would be the best choice for this sentence (reproduced below)?

   (2) <u>The RV</u> was a great purchase for our family and brought us all closer together.

   a. NO CHANGE
   b. Not surprisingly, the RV
   c. Furthermore, the RV
   d. As it turns out, the RV

3. Which of the following would be the best choice for this sentence (reproduced below)?

   Every morning (3) <u>we would wake up, eat breakfast, and broke camp.</u>

   a. NO CHANGE
   b. we would wake up, eat breakfast, and break camp.
   c. would we wake up, eat breakfast, and break camp?
   d. we are waking up, eating breakfast, and breaking camp.

4. Which of the following would be the best choice for this sentence (reproduced below)?

(4) We rejoiced as "hackers."

a. NO CHANGE
b. To a nagging problem of technology, we rejoiced as "hackers."
c. We rejoiced when we figured out how to "hack" a solution to a nagging technological problem.
d. To "hack" our way to a solution, we had to rejoice.

5. Which of the following would be the best choice for this sentence (reproduced below)?

(5) We even made some new friends in the process.

a. NO CHANGE
b. In the process was the friends we were making.
c. We are even making some new friends in the process.
d. We will make new friends in the process.

6. Which of the following would be the best choice for this sentence (reproduced below)?

(6) Above all, it allowed us to share adventures. While travelling across America, which we could not have experienced in cars and hotels.

a. NO CHANGE
b. Above all, it allowed us to share adventures while traveling across America
c. Above all, it allowed us to share adventures; while traveling across America
d. Above all, it allowed us to share adventures—while traveling across America

7. Which of the following would be the best choice for this sentence (reproduced below)?

(7) Those are also memories that my siblings and me have now shared with our own children.

a. NO CHANGE
b. Those are also memories that me and my siblings
c. Those are also memories that my siblings and I
d. Those are also memories that I and my siblings

Read the following section about Fred Hampton and answer Questions 8-20.

Fred Hampton desired to see lasting social change for African American people through nonviolent means and community recognition. (8) In the meantime, he became an African American activist during the American Civil Rights Movement and led the Chicago chapter of the Black Panther Party.

Hampton's Education

Hampton was born and raised (9) in Maywood of Chicago, Illinois in 1948. Gifted academically and a natural athlete, he became a stellar baseball player in high school. (10) After graduating from Proviso East High School in 1966, he later went on to study law at Triton Junior College.

While studying at Triton, Hampton joined and became a leader of the National Association for the Advancement of Colored People (NAACP). As a result of his leadership, the NAACP gained more than 500 members. Hampton worked relentlessly to acquire recreational facilities in the neighborhood and improve the educational resources provided to the impoverished black community of Maywood.

The Black Panthers

The Black Panther Party (BPP) (11) was another that formed around the same time as and was similar in function to the NAACP. Hampton was quickly attracted to the (12) Black Panther Party's approach to the fight for equal rights for African Americans. Hampton eventually joined the chapter and relocated to downtown Chicago to be closer to its headquarters.

His charismatic personality, organizational abilities, sheer determination, and rhetorical skills (13) enable him to quickly rise through the chapter's ranks. Hampton soon became the leader of the Chicago chapter of the BPP where he organized rallies, taught political education classes, and established a free medical clinic. (14) He also took part in the community police supervision project. He played an instrumental role in the BPP breakfast program for impoverished African American children.

Hampton's (15) greatest acheivement as the leader of the BPP may be his fight against street gang violence in Chicago. In 1969, (16) Hampton was held by a press conference where he made the gangs agree to a nonaggression pact known as the Rainbow Coalition. As a result of the pact, a multiracial alliance between blacks, Puerto Ricans, and poor youth was developed.

Assassination

(17) As the Black Panther Party's popularity and influence grew, the Federal Bureau of Investigation (FBI) placed the group under constant surveillance. In an attempt to neutralize the party, the FBI launched several harassment campaigns against the BPP, raided its headquarters in Chicago three times, and arrested over one hundred of the group's members. Hampton was shot during such a raid that occurred on the morning of December 4th 1969.

(18) In 1976; seven years after the event, it was revealed that William O'Neal, Hampton's trusted bodyguard, was an undercover FBI agent. (19) O'Neal will provide the FBI with detailed floor plans of the BPP's headquarters, identifying the exact location of Hampton's bed. It was because of these floor plans that the police were able to target and kill Hampton.

The assassination of Hampton fueled outrage amongst the African American community. It was not until years after the assassination that the police admitted wrongdoing. (20) The Chicago City Council now are commemorating December 4th as Fred Hampton Day.

8. Which of the following would be the best choice for this sentence (reproduced below)?

(8) <u>In the meantime,</u> he became an African American activist during the American Civil Rights Movement and led the Chicago chapter of the Black Panther Party.

    a. NO CHANGE
    b. Unfortunately,
    c. Finally,
    d. As a result,

9. Which of the following would be the best choice for this sentence (reproduced below)?

Hampton was born and raised (9) <u>in Maywood of Chicago, Illinois in 1948.</u>

    a. NO CHANGE
    b. in Maywood, of Chicago, Illinois in 1948.
    c. in Maywood of Chicago, Illinois, in 1948.
    d. in Chicago, Illinois of Maywood in 1948.

10. Which of these sentences, if any, should begin a new paragraph?
    a. There should be no new paragraph.
    b. After graduating from Proviso East High School in 1966, he later went on to study law at Triton Junior College.
    c. While studying at Triton, Hampton joined and became a leader of the National Association for the Advancement of Colored People (NAACP).
    d. As a result of his leadership, the NAACP gained more than 500 members.

11. Which of the following facts would be the most relevant to include here?
    a. NO CHANGE; best as written
    b. was another activist group that
    c. had a lot of members that
    d. was another school that

12. Which of the following would be the best choice for this sentence (reproduced below)?

Hampton was quickly attracted to the (12) <u>Black Panther Party's approach</u> to the fight for equal rights for African Americans.

    a. NO CHANGE
    b. Black Panther Parties approach
    c. Black Panther Partys' approach
    d. Black Panther Parties' approach

13. Which of the following would be the best choice for this sentence (reproduced below)?

His charismatic personality, organizational abilities, sheer determination, and rhetorical skills (13) <u>enable him to quickly rise</u> through the chapter's ranks.

a. NO CHANGE
b. are enabling him to quickly rise
c. enabled him to quickly rise
d. will enable him to quickly rise

14. Which of the following would be the best choice for this sentence (reproduced below)?

(14) <u>He also took part in the community police supervision project. He played an instrumental role</u> in the BPP breakfast program for impoverished African American children.

a. NO CHANGE
b. He also took part in the community police supervision project but played an instrumental role
c. He also took part in the community police supervision project, he played an instrumental role
d. He also took part in the community police supervision project and played an instrumental role

15. Which of these, if any, is misspelled?
a. None of these are misspelled.
b. greatest
c. acheivement
d. leader

16. Which of the following would be the best choice for this sentence (reproduced below)?

In 1969, (16) <u>Hampton was held by a press conference</u> where he made the gangs agree to a nonaggression pact known as the Rainbow Coalition.

a. NO CHANGE
b. Hampton held a press conference
c. Hampton, holding a press conference
d. Hampton to hold a press conference

17. Which of the following would be the best choice for this sentence (reproduced below)?

(17) <u>As the Black Panther Party's popularity and influence grew, the Federal Bureau of Investigation (FBI) placed the group under constant surveillance.</u>

a. NO CHANGE
b. The Federal Bureau of Investigation (FBI) placed the group under constant surveillance as the Black Panther Party's popularity and influence grew.
c. Placing the group under constant surveillance, the Black Panther Party's popularity and influence grew.
d. As their influence and popularity grew, the FBI placed the group under constant surveillance.

18. Which of the following would be the best choice for this sentence (reproduced below)?

(18) <u>In 1976; seven years after the event,</u> it was revealed that William O'Neal, Hampton's trusted bodyguard, was an undercover FBI agent.

a. NO CHANGE
b. In 1976, seven years after the event,
c. In 1976 seven years after the event,
d. In 1976. Seven years after the event,

19. Which of the following would be the best choice for this sentence (reproduced below)?

(19) <u>O'Neal will provide</u> the FBI with detailed floor plans of the BPP's headquarters, identifying the exact location of Hampton's bed.

a. NO CHANGE
b. O'Neal provides
c. O'Neal provided
d. O'Neal, providing

20. Which of the following would be the best choice for this sentence (reproduced below)?

(20) <u>The Chicago City Council now are commemorating December 4th as Fred Hampton Day.</u>

a. NO CHANGE
b. Fred Hampton Day by the Chicago City Council, December 4, is now commemorated.
c. Now commemorated December 4th is Fred Hampton Day.
d. The Chicago City Council now commemorates December 4th as Fred Hampton Day.

Read the essay entitled "Education is Essential to Civilization" and answer Questions 21-35.

Early in my career, (21) <u>a master's teacher shared this thought with me "Education is the last bastion of civility."</u> While I did not completely understand the scope of those words at the time, I have since come to realize the depth, breadth, truth, and significance of what he said. (22) <u>Education provides</u> society with a vehicle for (23) <u>raising it's children to be</u> civil, decent, human beings with something valuable to contribute to the world. It is really what makes us human and what (24) <u>distinguishes us as civilised creatures.</u>

Being "civilized" humans means being "whole" humans. Education must address the mind, body, and soul of students. (25) <u>It would be detrimental to society, only meeting the needs of the mind, if our schools were myopic in their focus.</u> As humans, we are multi-dimensional, multi-faceted beings who need more than head knowledge to survive. (26) <u>The human heart and psyche have to be fed in order for the mind to develop properly, and the body must be maintained and exercised to help fuel the working of the brain. Education is a basic human right, and it allows us to sustain a democratic society in which participation is fundamental to its success. It should inspire students to seek better solutions to world problems and to dream of a more equitable society.</u> Education should never discriminate on any basis, and it should create individuals who are self-sufficient, patriotic, and tolerant of (27) <u>others' ideas.</u>

(28) <u>All children can learn. Although not all children learn in the same manner.</u> All children learn best, however, when their basic physical needs are met and they feel safe, secure, and loved. Students are much more responsive to a teacher who values them and shows them respect as individual people. Teachers must model at all times the way they expect students to treat them and their peers. If teachers set high expectations for (29) <u>there students</u>, the students will rise to that high level. Teachers must make the well-being of students their primary focus and must not be afraid to let students learn from their own mistakes.

In the modern age of technology, a teacher's focus is no longer the "what" of the content, (30) <u>but more importantly, the 'why.'</u> Students are bombarded with information and have access to ANY information they need right at their fingertips. Teachers have to work harder than ever before to help students identify salient information (31) <u>so to think critically</u> about the information they encounter. Students have to (32) <u>read between the lines, identify bias, and determine</u> who they can trust in the milieu of ads, data, and texts presented to them.

Schools must work in consort with families in this important mission. While children spend most of their time in school, they are dramatically and indelibly shaped (33) <u>with the influences</u> of their family and culture. Teachers must not only respect this fact, (34) <u>but must strive</u> to include parents in the education of their children and must work to keep parents informed of progress and problems. Communication between classroom and home is essential for a child's success.

Humans have always aspired to be more, do more, and to better ourselves and our communities. This is where education lies, right at the heart of humanity's desire to be all that we can be. Education helps us strive for higher goals and better treatment of ourselves and others. I shudder to think what would become of us if education ceased to be the "last bastion of civility." (35) <u>We must be unapologetic about expecting excellence from our students? Our very existence depends upon it.</u>

21. Which of the following would be the best choice for this sentence (reproduced below)?

Early in my career, (21) <u>a master's teacher shared this thought with me "Education is the last bastion of civility."</u>

a. NO CHANGE
b. a master's teacher shared this thought with me: "Education is the last bastion of civility."
c. a master's teacher shared this thought with me: "Education is the last bastion of civility".
d. a master's teacher shared this thought with me. "Education is the last bastion of civility."

22. Which of the following would be the best choice for this sentence (reproduced below)?

(22) <u>Education provides</u> society with a vehicle for (23) <u>raising it's children to be</u> civil, decent, human beings with something valuable to contribute to the world.

a. NO CHANGE
b. Education provide
c. Education will provide
d. Education providing

23. Which of the following would be the best choice for this sentence (reproduced below)?

(22) <u>Education provides</u> society with a vehicle for (23) <u>raising it's children to be</u> civil, decent, human beings with something valuable to contribute to the world.

    a. NO CHANGE
    b. raises its children to be
    c. raising its' children to be
    d. raising its children to be

24. Which of these, if any, is misspelled?
    a. None of these are misspelled.
    b. distinguishes
    c. civilised
    d. creatures

25. Which of the following would be the best choice for this sentence (reproduced below)?

(25) <u>It would be detrimental to society, only meeting the needs of the mind, if our schools were myopic in their focus.</u>

    a. NO CHANGE
    b. It would be detrimental to society if our schools were myopic in their focus, only meeting the needs of the mind.
    c. Only meeting the needs of our mind, our schools were myopic in their focus, detrimental to society.
    d. Myopic is the focus of our schools, being detrimental to society for only meeting the needs of the mind.

26. Which of these sentences, if any, should begin a new paragraph?
    a. There should be no new paragraph.
    b. The human heart and psyche have to be fed in order for the mind to develop properly, and the body must be maintained and exercised to help fuel the working of the brain.
    c. Education is a basic human right, and it allows us to sustain a democratic society in which participation is fundamental to its success.
    d. It should inspire students to seek better solutions to world problems and to dream of a more equitable society.

27. Which of the following would be the best choice for this sentence (reproduced below)?

Education should never discriminate on any basis, and it should create individuals who are self-sufficient, patriotic, and tolerant of (27) <u>others' ideas.</u>

    a. NO CHANGE
    b. other's ideas
    c. others ideas
    d. others's ideas

28. Which of the following would be the best choice for this sentence (reproduced below)?

(28) All children can learn. Although not all children learn in the same manner.

a. NO CHANGE
b. All children can learn although not all children learn in the same manner.
c. All children can learn although, not all children learn in the same manner.
d. All children can learn, although not all children learn in the same manner.

29. Which of the following would be the best choice for this sentence (reproduced below)?

If teachers set high expectations for (29) there students, the students will rise to that high level.

a. NO CHANGE
b. they're students
c. their students
d. thare students

30. Which of the following would be the best choice for this sentence (reproduced below)?

In the modern age of technology, a teacher's focus is no longer the "what" of the content, (30) but more importantly, the 'why.'

a. NO CHANGE
b. but more importantly, the "why."
c. but more importantly, the 'why'.
d. but more importantly, the "why".

31. Which of the following would be the best choice for this sentence (reproduced below)?

Teachers have to work harder than ever before to help students identify salient information (31) so to think critically about the information they encounter.

a. NO CHANGE
b. and to think critically
c. but to think critically
d. nor to think critically

32. Which of the following would be the best choice for this sentence (reproduced below)?

Students have to (32) read between the lines, identify bias, and determine who they can trust in the milieu of ads, data, and texts presented to them.

a. NO CHANGE
b. read between the lines, identify bias, and determining
c. read between the lines, identifying bias, and determining
d. reads between the lines, identifies bias, and determines

33. Which of the following would be the best choice for this sentence (reproduced below)?

While children spend most of their time in school, they are dramatically and indelibly shaped (33) <u>with the influences</u> of their family and culture.

   a. NO CHANGE
   b. for the influences
   c. to the influences
   d. by the influences

34. Which of the following would be the best choice for this sentence (reproduced below)?

Teachers must not only respect this fact, (34) <u>but must strive</u> to include parents in the education of their children and must work to keep parents informed of progress and problems.

   a. NO CHANGE
   b. but to strive
   c. but striving
   d. but strived

35. Which of the following would be the best choice for this sentence (reproduced below)?

(35) <u>We must be unapologetic about expecting excellence from our students? Our very existence depends upon it.</u>

   a. NO CHANGE
   b. We must be unapologetic about expecting excellence from our students, our very existence depends upon it.
   c. We must be unapologetic about expecting excellence from our students—our very existence depends upon it.
   d. We must be unapologetic about expecting excellence from our students our very existence depends upon it.

Read the following passage and answer Questions 36-40.

Although many Missourians know that Harry S. Truman and Walt Disney hailed from their great state, probably far fewer know that it was also home to the remarkable George Washington Carver. (36) <u>As a child, George was driven to learn, and he loved painting.</u> At the end of the Civil War, Moses Carver, the slave owner who owned George's parents, decided to keep George and his brother and raise them on his farm.

He even went on to study art while in college but was encouraged to pursue botany instead. He spent much of his life helping others (37) <u>by showing them better ways to farm, his ideas improved agricultural productivity</u> in many countries. One of his most notable contributions to the newly emerging class of Black farmers was to teach them the negative effects of agricultural monoculture, i.e. (38) <u>growing the same crops in the same fields year after year, depleting the soil of much needed nutrients and results in a lesser yielding crop.</u>

Carver was an innovator, always thinking of new and better ways to do things, and is most famous for his over three hundred uses for the peanut. Toward the end of his career, (39) <u>Carver returns</u> to his first love of art. Through his artwork, he hoped to inspire people to see the beauty around them and to do great things themselves. (40) <u>Because Carver died,</u> he left his money to help fund ongoing agricultural research. Today, people still visit and study at the George Washington Carver Foundation at Tuskegee Institute.

36. Which of the following would be the best choice for this sentence?

(36) <u>As a child, George was driven to learn, and he loved painting.</u>

a. NO CHANGE
b. Move to the end of the first paragraph.
c. Move to the beginning of the first paragraph.
d. Move to the end of the second paragraph.

37. Which of the following would be the best choice for this sentence?

He spent much of his life helping others (37) <u>by showing them better ways to farm, his ideas improved agricultural productivity</u> in many countries.

a. NO CHANGE
b. by showing them better ways to farm his ideas improved agricultural productivity
c. by showing them better ways to farm . . . his ideas improved agricultural productivity
d. by showing them better ways to farm; his ideas improved agricultural productivity

38. Which of the following would be the best choice for this sentence?

(38) <u>growing the same crops in the same fields year after year, depleting the soil of much needed nutrients and results in a lesser yielding crop.</u>

a. NO CHANGE
b. growing the same crops in the same fields year after year, depleting the soil of much needed nutrients and resulting in a lesser yielding crop.
c. growing the same crops in the same fields year after year, depletes the soil of much needed nutrients and resulting in a lesser yielding crop.
d. grows the same crops in the same fields year after year, depletes the soil of much needed nutrients and resulting in a lesser yielding crop.

39. Which of the following would be the best choice for this sentence?

Toward the end of his career, (39) <u>Carver returns</u> to his first love of art.

a. NO CHANGE
b. Carver is returning
c. Carver returned
d. Carver was returning

40. Which of the following would be the best choice for this sentence?

(40) <u>Because Carver died,</u> he left his money to help fund ongoing agricultural research.

    a. NO CHANGE
    b. Although Carver died,
    c. When Carver died,
    d. Finally Carver died,

Read the following passage and answer Questions 41-44.

(41) <u>Christopher Columbus is often credited for discovering America. This is incorrect.</u> First, it is impossible to "discover" something where people already live; however, Christopher Columbus did explore places in the New World that were previously untouched by Europe, so the term "explorer" would be more accurate. Another correction must be made, as well: Christopher Columbus was not the first European explorer to reach the present day Americas! Rather, it was Leif Erikson who first came to the New World and contacted the natives, nearly five hundred years before Christopher Columbus.

Leif Erikson, the son of Erik the Red (a famous Viking outlaw and explorer in his own right), was born in either (42) <u>970 or 980. Depending on which historian you seek</u>. His own family, though, did not raise Leif, which was a Viking tradition. Instead, one of Erik's prisoners taught Leif reading and writing, languages, sailing, and weaponry. At age 12, Leif was considered a man and returned to his family. He killed a man during a dispute shortly after his return, and the council banished the Erikson clan to Greenland. In 999, Leif left Greenland and traveled to Norway where he would serve as a guard to King Olaf Tryggvason. It was there that he became a convert to Christianity. Leif later tried to return home with the intention of taking supplies and spreading Christianity to Greenland, however his ship was blown off course and he arrived in a strange new land: present day Newfoundland, Canada.

When he finally returned to his adopted homeland, Greenland, (43) <u>Leif consults</u> with a merchant who had also seen the shores of this previously unknown land we now know as Canada. The son of the legendary Viking explorer then gathered a crew of 35 men and set sail. Leif became the first European to touch foot in the New World as he explored present-day Baffin Island and Labrador, Canada. His crew called the land Vinland since it was plentiful with grapes.

During their time in present-day Newfoundland, Leif's expedition made contact with the natives whom they referred to as Skraelings (which translates to "wretched ones" in Norse). There are several secondhand accounts of their meetings. Some contemporaries described trade between the peoples. (44) <u>Other accounts describes</u> clashes where the Skraelings defeated the Viking explorers with long spears, while still others claim the Vikings dominated the natives. Regardless of the circumstances, it seems that the Vikings made contact of some kind. This happened around 1000, nearly five hundred years before Columbus famously sailed the ocean blue.

41. Which of the following would be the best choice for this sentence?

(41) <u>Christopher Columbus is often credited for discovering America. This is incorrect.</u>

a. NO CHANGE
b. Christopher Columbus is often credited for discovering America this is incorrect.
c. Christopher Columbus is often credited for discovering America, this is incorrect.
d. Christopher Columbus is often credited for discovering America: this is incorrect.

42. Which of the following would be the best choice for this sentence?

Leif Erikson, the son of Erik the Red (a famous Viking outlaw and explorer in his own right), was born in either (42) <u>970 or 980.</u>

a. NO CHANGE
b. 970 or 980! depending on which historian you seek.
c. 970 or 980, depending on which historian you seek.
d. 970 or 980; depending on which historian you seek.

43. Which of the following would be the best choice for this sentence?

When he finally returned to his adopted homeland, Greenland, (43) <u>Leif consults</u> with a merchant who had also seen the shores of this previously unknown land we now know as Canada.

a. NO CHANGE
b. Leif consulted
c. Leif consulting
d. Leif was consulted

44. Which of the following would be the best choice for this sentence?

(44) <u>Other accounts describes</u> clashes where the Skraelings defeated the Viking explorers with long spears, while still others claim the Vikings dominated the natives.

a. NO CHANGE
b. Other account's describe
c. Other accounts describe
d. Others account's describes

Math

1. Which of the following inequalities is equivalent to $3 - \frac{1}{2}x \geq 2$?
   a. $x \geq 2$
   b. $x \leq 2$
   c. $x \geq 1$
   d. $x \leq 1$

2. If $g(x) = x^3 - 3x^2 - 2x + 6$ and $f(x) = 2$, then what is $g(f(x))$?
   a. -26
   b. 6
   c. $2x^3 - 6x^2 - 4x + 12$
   d. -2

3. The graph of which function has an x-intercept of -2?
   a. $y = 2x - 3$
   b. $y = 4x + 2$
   c. $y = x^2 + 5x + 6$
   d. $y = -\frac{1}{2} \times 2^x$

4. The table below displays the number of three-year-olds at Kids First Daycare who are potty-trained and those who still wear diapers.

	Potty-trained	Wear diapers	Total
Boys	26	22	48
Girls	34	18	52
Total	60	40	

What is the probability that a three-year-old girl chosen at random from the school is potty-trained?
   a. 52 percent
   b. 34 percent
   c. 65 percent
   d. 57 percent

5. What is the volume of a cube with the side equal to 3 inches?
   a. 6 in³
   b. 27 in³
   c. 9 in³
   d. 3 in³

6. What is the volume of a rectangular prism with the height of 3 centimeters, a width of 5 centimeters, and a depth of 11 centimeters?
   a. 19 cm³
   b. 165 cm³
   c. 225 cm³
   d. 150 cm³

7. What is the volume of a cylinder, in terms of π, with a radius of 5 inches and a height of 10 inches?
   a. 250 π in³
   b. 50 π in³
   c. 100 π in³
   d. 200 π in³

8. What is the solution to the following system of equations?
$$x^2 - 2x + y = 8$$
$$x - y = -2$$

a. $(-2, 3)$
b. There is no solution.
c. $(-2, 0)\ (1, 3)$
d. $(-2, 0)\ (3, 5)$

9. An equation for the line passing through the origin and the point $(2, 1)$ is
a. $y = 2x$
b. $y = \frac{1}{2}x$
c. $y = x - 2$
d. $y = x - 1$

10. A rectangle was formed out of pipe cleaner. Its length was $\frac{1}{2}$ feet, and its width was $\frac{11}{2}$ inches. What is its area in square inches?

a. $\frac{11}{4}$ inch^2

b. $\frac{11}{2}$ inch^2

c. 22 inch^2

d. 33 inch^2

11. What type of function is modeled by the values in the following table?

X	f(x)
1	2
2	4
3	8
4	16
5	32

a. Linear
b. Exponential
c. Quadratic
d. Cubic

12. Two cards are drawn from a shuffled deck of 52 cards. What's the probability that both cards are Kings if the first card isn't replaced after it's drawn?

a. $\frac{1}{169}$

b. $\frac{1}{221}$

c. $\frac{1}{13}$

d. $\frac{4}{13}$

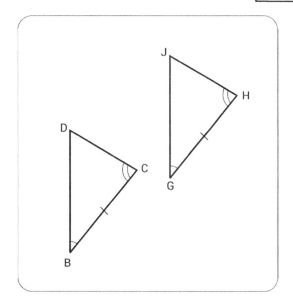

13. In the image above, what is demonstrated by the two triangles?
    a. According to Side-Side-Side, the triangles are congruent.
    b. According to Angle-Angle-Angle, the triangles are congruent.
    c. According to Angle-Side-Angle, the triangles are congruent.
    d. There is not enough information to prove the two triangles are congruent.

14. Write the expression for six less than three times the sum of twice a number and one.
    a. $2x + 1 - 6$
    b. $3x + 1 - 6$
    c. $3(x + 1) - 6$
    d. $3(2x + 1) - 6$

15. Five of six numbers have a sum of 25. The average of all six numbers is 6. What is the sixth number?
    a. 8
    b. 10
    c. 11
    d. 12

16. $(2x - 4y)^2 =$
    a. $4x^2 - 16xy + 16y^2$
    b. $4x^2 - 8xy + 16y^2$
    c. $4x^2 - 16xy - 16y^2$
    d. $2x^2 - 8xy + 8y^2$

17. What are the zeros of $f(x) = x^2 + 4$?
    a. $x = -4$
    b. $x = \pm 2i$
    c. $x = \pm 2$
    d. $x = \pm 4i$

18. Which of the following shows the correct result of simplifying the following expression:
$$(7n + 3n^3 + 3) + (8n + 5n^3 + 2n^4)$$
a. $9n^4 + 15n - 2$
b. $2n^4 + 5n^3 + 15n - 2$
c. $9n^4 + 8n^3 + 15n$
d. $2n^4 + 8n^3 + 15n + 3$

19. What is the slope of this line?

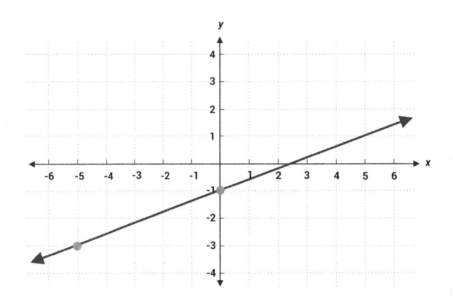

a. 2

b. $\frac{5}{2}$

c. $\frac{1}{2}$

d. $\frac{2}{5}$

20. What is the product of the following expression?
$$(4x - 8)(5x^2 + x + 6)$$
a. $20x^3 - 36x^2 + 16x - 48$
b. $6x^3 - 41x^2 + 12x + 15$
c. $204 + 11x^2 - 37x - 12$
d. $2x^3 - 11x^2 - 32x + 20$

21. What is the solution for the following equation?
$$\frac{x^2 + x - 30}{x - 5} = 11$$
a. $x = -6$
b. There is no solution.
c. $x = 16$
d. $x = 5$

22. If x is not zero, then $\frac{3}{x} + \frac{5u}{2x} - \frac{u}{4} =$

a. $\frac{12+10u-ux}{4x}$

b. $\frac{3+5u-ux}{x}$

c. $\frac{12x+10u+ux}{4x}$

d. $\frac{12+10u-u}{4x}$

23. What are the zeros of the function: $f(x) = x^3 + 4x^2 + 4x$?

a. -2
b. 0, -2
c. 2
d. 0, 2

24. Is the following function even, odd, neither, or both?

$$y = \frac{1}{2}x^4 + 2x^2 - 6$$

a. Even
b. Odd
c. Neither
d. Both

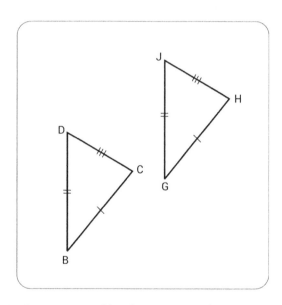

25. In the image above, what is demonstrated by the two triangles?
a. According to Side-Side-Side, the triangles are congruent.
b. According to Angle-Angle-Angle, the triangles are congruent.
c. According to Angle-Side-Angle, the triangles are congruent.
d. There is not enough information to prove the two triangles are congruent.

26. $Sin(x) = -0.8$. If x and y are complementary, what is the cos (y)?
    a. 0.2
    b. -0.2
    c. -0.8
    d. 0.8

27. Given the following triangle, what's the length of the missing side? Round the answer to the nearest tenth.

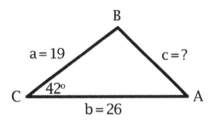

    a. 17.0
    b. 17.4
    c. 18.0
    d. 18.4

28. For the following similar triangles, what are the values of x and y (rounded to one decimal place)?

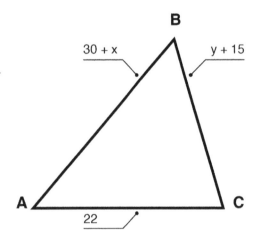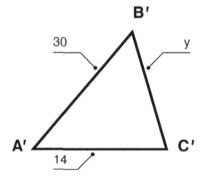

    a. $x = 16.5, y = 25.1$
    b. $x = 19.5, y = 24.1$
    c. $x = 17.1, y = 26.3$
    d. $x = 26.3, y = 17.1$

29. What are the center and radius of a circle with equation $4x^2 + 4y^2 - 16x - 24y + 51 = 0$?
    a. Center (3, 2) and radius ½
    b. Center (2, 3) and radius ½
    c. Center (3, 2) and radius ¼
    d. Center (2, 3) and radius ¼

30.

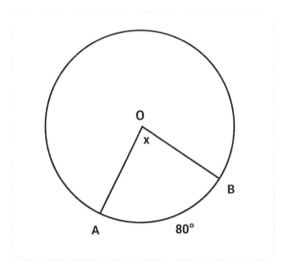

The area of circle O is 49π m. What is the area of the sector formed by $\angle AOB$?
- a. 80π m
- b. 10.9π m
- c. 4.9π m
- d. 10π m

31. Where does the point (-3, -4) lie on the circle with the equation $(x)^2 + (y)^2 = 25$?
- a. Inside of the circle.
- b. Outside of the circle.
- c. On the circle.
- d. There is not enough information to tell.

32. What is the length of the other leg of a right triangle with a hypotenuse of 10 inches and a leg of 8 inches?
- a. 6 in
- b. 18 in
- c. 80 in
- d. 13 in

33. What's the probability of rolling a 6 at least once in two rolls of a die?
- a. $\frac{1}{3}$
- b. $\frac{1}{36}$
- c. $\frac{1}{6}$
- d. $\frac{11}{36}$

34. Karen gets paid a weekly salary and a commission for every sale that she makes. The table below shows the number of sales and her pay for different weeks.

Sales	2	7	4	8
Pay	$380	$580	$460	$620

Which of the following equations represents Karen's weekly pay?

a. $y = 90x + 200$
b. $y = 90x - 200$
c. $y = 40x + 300$
d. $y = 40x - 300$

35. The square and circle have the same center. The circle has a radius of r. What is the area of the shaded region?

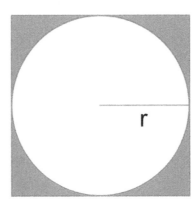

a. $r^2 - \pi r^2$
b. $4r^2 - 2\pi r$
c. $(4 - \pi)r^2$
d. $(\pi - 1)r^2$

36. The graph shows the position of a car over a 10-second time interval. Which of the following is the correct interpretation of the graph for the interval 1 to 3 seconds?

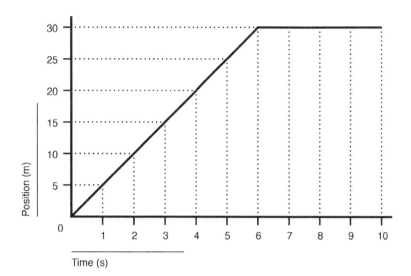

Time (s)

a. The car remains in the same position.
b. The car is traveling at a speed of 5m/s.
c. The car is traveling up a hill.
d. The car is traveling at 5mph.

37. Which of the ordered pairs below is a solution to the following system of inequalities?
$$y > 2x - 3$$
$$y < -4x + 8$$

a. $(4, 5)$
b. $(-3, -2)$
c. $(3, -1)$
d. $(5, 2)$

38. Which equation best represents the scatterplot below?

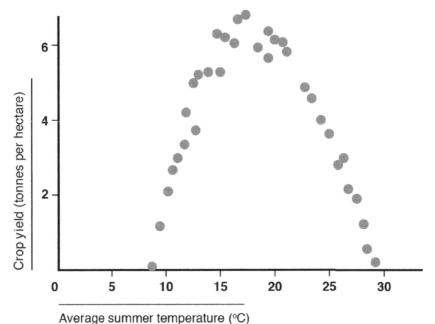

a. $y = 3x - 4$

b. $y = 2x^2 + 7x - 9$

c. $y = (3)(4^x)$

d. $y = -\frac{1}{14}x^2 + 2x - 8$

39. Using trigonometric ratios for a right angle, what is the value of the closest angle whose adjacent side is equal to 7.071 centimeters and whose hypotenuse is equal to 10 centimeters?
   a. 15°
   b. 30°
   c. 45°
   d. 90°

40. Using trigonometric ratios, what is the value of the other angle whose opposite side is equal to 1 inch and whose adjacent side is equal to the square root of 3 inches?
   a. 15°
   b. 30°
   c. 45°
   d. 90°

41. Suppose an investor deposits $1,200 into a bank account that accrues 1 percent interest per month. Assuming x represents the number of months since the deposit and y represents the money in the account, which of the following exponential functions models the scenario?
   a. $y = (0.01)(1200^x)$
   b. $y = (1200)(0.01^x)$
   c. $y = (1.01)(1200^x)$
   d. $y = (1200)(1.01^x)$

42. Solve for x: $\frac{2x}{5} - 1 = 59$.

    a. 60
    b. 145
    c. 150
    d. 115

43. Which of the following is the result of simplifying the expression:

$$\frac{4a^{-1}b^3}{a^4b^{-2}} \times \frac{3a}{b}$$

    a. $12a^3b^5$

    b. $12\frac{b^4}{a^4}$

    c. $\frac{12}{a^4}$

    d. $7\frac{b^4}{a}$

44. What is the simplified quotient of the following equation?

$$\frac{5x^3}{3x^2y} \div \frac{25}{3y^9}$$

    a. $\frac{125x}{9y^{10}}$

    b. $\frac{x}{5y^8}$

    c. $\frac{5}{xy^8}$

    d. $\frac{xy^8}{5}$

45. What is the solution for the equation $\tan(x) + 1 = 0$, where $0 \le x < 2\pi$?

    a. $x = \frac{3\pi}{4}, \frac{5\pi}{4}$

    b. $x = \frac{3\pi}{4}, \frac{\pi}{4}$

    c. $x = \frac{5\pi}{4}, \frac{7\pi}{4}$

    d. $x = \frac{3\pi}{4}, \frac{7\pi}{4}$

No Calculator Questions

46. In Jim's school, there are 3 girls for every 2 boys. There are 650 students in total. Using this information, how many students are girls?

    a. 260
    b. 130
    c. 65
    d. 390

47. Two chords intersect inside of a circle. The segments of one chord have the lengths 7 and $2x + 2$. The segments of the other chord have lengths 6 and $x + 5$. What are the lengths of these chords?
    a. 13 units
    b. 2 units
    c. 42 units
    d. 6 units

48. At the store, Jan spends $90 on apples and oranges. Apples cost $1 each and oranges cost $2 each. If Jan buys the same number of apples as oranges, how many oranges did she buy?
    a. 20
    b. 25
    c. 30
    d. 35

49. A train traveling 50 miles per hour takes a trip lasting 3 hours. If a map has a scale of 1 inch per 10 miles, how many inches apart are the train's starting point and ending point on the map?
    a. 14
    b. 12
    c. 13
    d. 15

50. A traveler takes an hour to drive to a museum, spends 3 hours and 30 minutes there, and takes half an hour to drive home. What percentage of his or her time was spent driving?
    a. 15%
    b. 30%
    c. 40%
    d. 60%

51. A truck is carrying three cylindrical barrels. Their bases have a diameter of 2 feet, and they have a height of 3 feet. What is the total volume of the three barrels in cubic feet?
    a. 3π
    b. 9π
    c. 12π
    d. 15π

52. The area of a given rectangle is 24 square centimeters. If the measure of each side is multiplied by 3, what is the area of the new figure?
    a. 48 cm^2
    b. 72 cm^2
    c. 216 cm^2
    d. 13,824 cm^2

53. If $3x = 6y = -2z = 24$, then what does $4xy + z$ equal?

54. If $4x - 3 = 5$, then $x =$

55. What is the solution to $4 \times 7 + (25 - 21)^2 \div 2$?

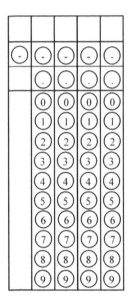

56. Solve the following:

$$\left(\sqrt{36} \times \sqrt{16}\right) - 3^2$$

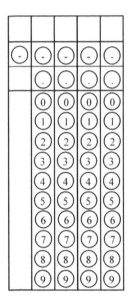

57. What is the overall median of Dwayne's current test scores: 78, 92, 83, 97?

58. What is the value of $x^2 - 2xy + 2y^2$ when $x = 2, y = 3$?

Answer Explanations for Practice Test #2

Reading

1. C: Gulliver becomes acquainted with the people and practices of his new surroundings. Choice *C* is the correct answer because it most extensively summarizes the entire passage. While Choices *A* and *B* are reasonable possibilities, they reference portions of Gulliver's experiences, not the whole. Choice *D* is incorrect because Gulliver doesn't express repentance or sorrow in this particular passage.

2. A: Principal refers to *chief* or *primary* within the context of this text. Choice *A* is the answer that most closely aligns with this definition. Choices *B* and *D* make reference to a helper or followers while Choice *C* doesn't meet the description of Reldresal from the passage.

3. C: One can reasonably infer that Gulliver is considerably larger than the children who were playing around him because multiple children could fit into his hand. Choice *A* is incorrect because there is no indication of stress in Gulliver's tone. Choices *B* and *D* aren't the best answers because though Gulliver seems fond of his new acquaintances, he didn't travel there with the intentions of meeting new people or to express a definite love for them in this particular portion of the text.

4. C: The emperor made a *definitive decision* to expose Gulliver to their native customs. In this instance, the word *mind* was not related to a vote, question, or cognitive ability.

5. A: Choice *A* is correct. This assertion does *not* support the fact that games are a commonplace event in this culture because it mentions conduct, not games. Choices *B*, *C*, and *D* are incorrect because these do support the fact that games were a commonplace event.

6. B: Choice *B* is the only option that mentions the correlation between physical ability and leadership positions. Choices *A* and *D* are unrelated to physical strength and leadership abilities. Choice *C* does not make a deduction that would lead to the correct answer—it only comments upon the abilities of common townspeople.

7. D: It emphasizes Mr. Utterson's anguish in failing to identify Hyde's whereabouts. Context clues indicate that Choice *D* is correct because the passage provides great detail of Mr. Utterson's feelings about locating Hyde. Choice *A* does not fit because there is no mention of Mr. Lanyon's mental state. Choice *B* is incorrect; although the text does make mention of bells, Choice *B* is not the *best* answer overall. Choice *C* is incorrect because the passage clearly states that Mr. Utterson was determined, not unsure.

8. A: In the city. The word *city* appears in the passage several times, thus establishing the location for the reader.

9. B: It scares children. The passage states that the Juggernaut causes the children to scream. Choices *A* and *D* don't apply because the text doesn't mention either of these instances specifically. Choice *C* is incorrect because there is nothing in the text that mentions space travel.

10. B: To constantly visit. The mention of *morning, noon,* and *night* make it clear that the word *haunt* refers to frequent appearances at various locations. Choice *A* doesn't work because the text makes no mention of levitating. Choices *C* and *D* are not correct because the text makes mention of Mr. Utterson's anguish and disheartenment because of his failure to find Hyde but does not make mention of Mr. Utterson's feelings negatively affecting anyone else.

11. D: This is an example of alliteration. Choice *D* is the correct answer because of the repetition of the *L*-words. Hyperbole is an exaggeration, so Choice *A* doesn't work. No comparison is being made, so no simile or metaphor is being used, thus eliminating Choices *B* and *C*.

12. D: The speaker intends to continue to look for Hyde. Choices *A* and *B* are not possible answers because the text doesn't refer to any name changes or an identity crisis, despite Mr. Utterson's extreme obsession with finding Hyde. The text also makes no mention of a mistaken identity when referring to Hyde, so Choice *C* is also incorrect.

13. A: The tone is exasperated. While contemplative is an option because of the inquisitive nature of the text, Choice *A* is correct because the speaker is frustrated by the thought of being included when he felt that the fellow members of his race were being excluded. The speaker is not nonchalant, nor accepting of the circumstances which he describes.

14. C: Choice *C, contented,* is the only word that has different meaning. Furthermore, the speaker expresses objection and disdain throughout the entire text.

15. B: To address the hypocrisy of the Fourth of July holiday. While the speaker makes biblical references, it is not the main focus of the passage, thus eliminating Choice *A* as an answer. The passage also makes no mention of wealthy landowners and doesn't speak of any positive response to the historical events, so Choices *C* and *D* are not correct.

16. D: Choice *D* is the correct answer because it clearly makes reference to justice being denied.

17. D: Hyperbole. Choices *A* and *B* are unrelated. Assonance is the repetition of sounds and commonly occurs in poetry. Parallelism refers to two statements that correlate in some manner. Choice *C* is incorrect because amplification normally refers to clarification of meaning by broadening the sentence structure, while hyperbole refers to a phrase or statement that is being exaggerated.

18. C: Choice *C* is correct because the speaker is clear about his intention and stance throughout the text; thus, it's not true that he makes biblical references to display his own equivocation and that of those that he represents. Choice *A* could be true, but the words "common text" is arguable because not everyone will understand the reference. Choice *B* is also partially true, as another group of people affected by slavery are being referenced. However, the speaker is not trying to convince the audience that injustices have been committed, as it is already understood there have been injustices committed. Choice *D* is also close to the correct answer, but it is not the best answer choice possible.

19. B: A period of time. It is apparent that Lincoln is referring to a period of time within the context of the passage because of how the sentence is structured with the word *ago.*

20. C: Lincoln's reference to *the brave men, living and dead, who struggled here,* proves that he is referring to a battlefield. Choices *A* and *B* are incorrect, as a *civil war* is mentioned and not a war with France or a war in the Sahara Desert. Choice *D* is incorrect because it does not make sense to consecrate a President's ground instead of a battlefield ground for soldiers who died during the American Civil War.

21. D: Abraham Lincoln is the former president of the United States, and he references a "civil war" during his address.

22. A: The audience should perpetuate the ideals of freedom that the soldiers died fighting for. Lincoln doesn't address any of the topics outlined in Choices *B*, *C*, or *D*. Therefore, Choice *A* is the correct answer.

23. D: Choice *D* is the correct answer because of the repetition of the word *people* at the end of the passage. Choice A, antimetabole, is the repetition of words in a phrase or clause but in reverse order, such as: "I do what I like, and like what I do." Choice *B*, *antiphrasis*, is a form of denial of an assertion in a text. Choice *C*, *anaphora*, is the repetition that occurs at the beginning of sentences.

24. A: Choice *A* is correct because Lincoln's intention was to memorialize the soldiers who had fallen as a result of war as well as celebrate those who had put their lives in danger for the sake of their country. Choices *B* and *D* are incorrect because Lincoln's speech was supposed to foster a sense of pride among the members of the audience while connecting them to the soldiers' experiences.

25. A: The word *patronage* most nearly means *auspices*, which means *protection* or *support*. Choice *B*, *aberration*, means *deformity* and does not make sense within the context of the sentence. Choice *C*, *acerbic,* means *bitter* and also does not make sense in the sentence. Choice *D*, *adulation*, is a positive word meaning *praise*, and thus does not fit with the word *condescending* in the sentence.

26. D: *Working man* is most closely aligned with Choice *D*, *bourgeois.* In the context of the speech, the word *bourgeois* means *working* or *middle class*. Choice *A*, *Plebeian*, does suggest *common people*; however, this is a term that is specific to ancient Rome. Choice *B*, *viscount*, is a European title used to describe a specific degree of nobility. Choice *C*, *entrepreneur*, is a person who operates their own business.

27. C: In the context of the speech, the term *working man* most closely correlates with Choice *C*, *working man is someone who works for wages among the middle class.* Choice *A* is not mentioned in the passage and is off-topic. Choice *B* may be true in some cases, but it does not reflect the sentiment described for the term *working man* in the passage. Choice *D* may also be arguably true. However, it is not given as a definition but as *acts* of the working man, and the topics of *field, factory,* and *screen* are not mentioned in the passage.

28. D: *Enterprise* most closely means *cause*. Choices *A, B,* and *C* are all related to the term *enterprise*. However, Dickens speaks of a *cause* here, not a company, courage, or a game. *He will stand by such an enterprise* is a call to stand by a cause to enable the working man to have a certain autonomy over his own economic standing. The very first paragraph ends with the statement that the working man *shall . . . have a share in the management of an institution which is designed for his benefit.*

29. B: The speaker's salutation is one from an entertainer to his audience and uses the friendly language to connect to his audience before a serious speech. Recall in the first paragraph that the speaker is there to "accompany [the audience] . . . through one of my little Christmas books," making him an author there to entertain the crowd with his own writing. The speech preceding the reading is the passage itself, and, as the tone indicates, a serious speech addressing the "working man." Although the passage speaks of employers and employees, the speaker himself is not an employer of the audience, so Choice *A* is incorrect. Choice *C* is also incorrect, as the salutation is not used ironically, but sincerely, as the speech addresses the wellbeing of the crowd. Choice *D* is incorrect because the speech is not given by a politician, but by a writer.

30. B: For the working man to have a say in his institution which is designed for his benefit. Choice *A* is incorrect because that is the speaker's *first* desire, not his second. Choices *C* and *D* are tricky because the language of both of these is mentioned after the word *second*. However, the speaker doesn't get to the second wish until the next sentence. Choices *C* and *D* are merely prepositions preparing for the statement of the main clause, Choice *B*.

31. D: The use of "I" could serve to have a "hedging" effect, allow the reader to connect with the author in a more personal way, and cause the reader to empathize more with the egrets. However, it doesn't distance the reader from the text, making Choice *D* the answer to this question.

32. C: The quote provides an example of a warden protecting one of the colonies. Choice *A* is incorrect because the speaker of the quote is a warden, not a hunter. Choice *B* is incorrect because the quote does not lighten the mood, but shows the danger of the situation between the wardens and the hunters. Choice *D* is incorrect because there is no humor found in the quote.

33. D: A *rookery* is a colony of breeding birds. Although *rookery* could mean Choice *A*, houses in a slum area, it does not make sense in this context. Choices *B* and *C* are both incorrect, as this is not a place for hunters to trade tools or for wardens to trade stories.

34. B: An important bird colony. The previous sentence is describing "twenty colonies" of birds, so what follows should be a bird colony. Choice *A* may be true, but we have no evidence of this in the text. Choice *C* does touch on the tension between the hunters and wardens, but there is no official "Bird Island Battle" mentioned in the text. Choice *D* does not exist in the text.

35. D: To demonstrate the success of the protective work of the Audubon Association. The text mentions several different times how and why the association has been successful and gives examples to back this fact. Choice *A* is incorrect because although the article, in some instances, calls certain people to act, it is not the purpose of the entire passage. There is no way to tell if Choices *B* and *C* are correct, as they are not mentioned in the text.

36. C: To have a better opportunity to hunt the birds. Choice *A* might be true in a general sense, but it is not relevant to the context of the text. Choice *B* is incorrect because the hunters are not studying lines of flight to help wardens, but to hunt birds. Choice *D* is incorrect because nothing in the text mentions that hunters are trying to build homes underneath lines of flight of birds for good luck.

37. A: It introduces certain insects that transition from water to air. Choice *B* is incorrect because although the passage talks about gills, it is not the central idea of the passage. Choices *C* and *D* are incorrect because the passage does not "define" or "invite," but only serves as an introduction to stoneflies, dragonflies, and mayflies and their transition from water to air.

38. C: The act of shedding part or all of the outer shell. Choices *A*, *B*, and *D* are incorrect.

39. B: The first paragraph serves as a contrast to the second. Notice how the first paragraph goes into detail describing how insects are able to breathe air. The second paragraph acts as a contrast to the first by stating "[i]t is of great interest to find that, nevertheless, a number of insects spend much of their time under water." Watch for transition words such as "nevertheless" to help find what type of passage you're dealing with.

40: C: The stage of preparation in between molting is acted out in the water, while the last stage is in the air. Choices *A, B,* and *D* are all incorrect. *Instars* is the phase between two periods of molting, and the text explains when these transitions occur.

41. C: The author's tone is informative and exhibits interest in the subject of the study. Overall, the author presents us with information on the subject. One moment where personal interest is depicted is when the author states, "It is of great interest to find that, nevertheless, a number of insects spend much of their time under water."

42. C: Their larva can breathe the air dissolved in water through gills of some kind. This is stated in the last paragraph. Choice *A* is incorrect because the text mentions this in a general way at the beginning of the passage concerning "insects as a whole." Choice *B* is incorrect because this is stated of beetles and water-bugs, and not the insects in question. Choice *D* is incorrect because this is the opposite of what the text says of instars.

43. D: To enlighten the audience on the habits of sun-fish and their hatcheries. Choice *A* is incorrect because although the Adirondack region is mentioned in the text, there is no cause or effect relationships between the region and fish hatcheries depicted here. Choice *B* is incorrect because the text does not have an agenda, but rather is meant to inform the audience. Finally, Choice *C* is incorrect because the text says nothing of how sun-fish mate.

44. B: The word *wise* in this passage most closely means *manner*. Choices *A* and *C* are synonyms of *wise*; however, they are not relevant in the context of the text. Choice *D, ignorance,* is opposite of the word *wise* and is therefore incorrect.

45. A: Fish at the stage of development where they are capable of feeding themselves. Even if the word *fry* isn't immediately known to the reader, the context gives a hint when it says "until the fry are hatched out and are sufficiently large to take charge of themselves."

46. B: The sun-fish builds it with her tail and snout. The text explains this in the second paragraph: "she builds, with her tail and snout, a circular embankment 3 inches in height and 2 thick." Choice *A* is used in the text as a simile.

47. D: To conclude a sequence and add a final detail. The concluding sequence is expressed in the phrase "[t]he mother sun-fish, having now built or provided her 'hatchery.'" The final detail is the way in which the sun-fish guards the "inclosure." Choices *A, B,* and *C* are incorrect.

Writing

1. B: Move the sentence so that it comes before the preceding sentence. For this question, place the underlined sentence in each prospective choice's position. To keep as-is is incorrect because the father "going crazy" doesn't logically follow the fact that he was a "city slicker." Choice *C* is incorrect because the sentence in question is not a concluding sentence and does not transition smoothly into the second paragraph. Choice *D* is incorrect because the sentence doesn't necessarily need to be omitted since it logically follows the very first sentence in the passage.

2. D: Choice *D* is correct because "As it turns out" indicates a contrast from the previous sentiment, that the RV was a great purchase. Choice *A* is incorrect because the sentence needs an effective transition from the paragraph before. Choice *B* is incorrect because the text indicates it *is* surprising that the RV

was a great purchase because the author was skeptical beforehand. Choice *C* is incorrect because the transition "Furthermore" does not indicate a contrast.

3. B: This sentence calls for parallel structure. Choice *B* is correct because the verbs "wake," "eat," and "break" are consistent in tense and parts of speech. Choice *A* is incorrect because the words "wake" and "eat" are present tense while the word "broke" is in past tense. Choice *C* is incorrect because this turns the sentence into a question, which doesn't make sense within the context. Choice *D* is incorrect because it breaks tense with the rest of the passage. "Waking," "eating," and "breaking" are all present participles, and the context around the sentence is in past tense.

4. C: Choice *C* is correct because it is clear and fits within the context of the passage. Choice *A* is incorrect because "We rejoiced as 'hackers'" does not give a reason *why* hacking was rejoiced. Choice *B* is incorrect because it does not mention a solution being found and is therefore not specific enough. Choice *D* is incorrect because the meaning is eschewed by the helping verb "had to rejoice," and the sentence does not give enough detail as to what the problem entails.

5. A: The original sentence is correct because the verb tense as well as the meaning aligns with the rest of the passage. Choice *B* is incorrect because the order of the words makes the sentence more confusing than it otherwise would be. Choice *C* is incorrect because "We are even making" is in present tense. Choice *D* is incorrect because "We will make" is future tense. The surrounding text of the sentence is in past tense.

6. B: Choice *B* is correct because there is no punctuation needed if a dependent clause ("while traveling across America") is located behind the independent clause ("it allowed us to share adventures"). Choice *A* is incorrect because there are two dependent clauses connected and no independent clause, and a complete sentence requires at least one independent clause. Choice *C* is incorrect because of the same reason as Choice *A*. Semicolons have the same function as periods: there must be an independent clause on either side of the semicolon. Choice *D* is incorrect because the dash simply interrupts the complete sentence.

7. C: The rules for "me" and "I" is that one should use "I" when it is the subject pronoun of a sentence, and "me" when it is the object pronoun of the sentence. Break the sentence up to see if "I" or "me" should be used. To say "Those are memories that I have now shared" is correct, rather than "Those are memories that me have now shared." Choice *D* is incorrect because "my siblings" should come before "I."

8. D: Choice *D* is correct because Fred Hampton becoming an activist was a direct result of him wanting to see lasting social change for Black people. Choice *A* doesn't make sense because "In the meantime" denotes something happening at the same time as another thing. Choice *B* is incorrect because the text's tone does not indicate that becoming a civil rights activist is an unfortunate path. Choice *C* is incorrect because "Finally" indicates something that comes last in a series of events, and the word in question is at the beginning of the introductory paragraph.

9. A: Choice *A* is correct because there should be a comma between the city and state. Choice *B* is incorrect because the comma after "Maywood" interrupts the phrase "Maywood of Chicago." Choice *C* is incorrect because a comma after "Illinois" is unnecessary. Finally, Choice *D* is incorrect because the order of the sentence designates that Chicago, Illinois is in Maywood, which is incorrect.

10. C: This is a difficult question. The paragraph is incorrect as-is because it is too long and thus loses the reader halfway through. Choice *C* is correct because if the new paragraph began with "While studying at

Triton," we would see a smooth transition from one paragraph to the next. We can also see how the two paragraphs are logically split in two. The first half of the paragraph talks about where he studied. The second half of the paragraph talks about the NAACP and the result of his leadership in the association. If we look at the passage as a whole, we can see that there are two main topics that should be broken into two separate paragraphs.

11. B: The BPP is another activist group. We can figure out this answer by looking at context clues. We know that the BPP is "similar in function" to the NAACP. To find out what the NAACP's function is, we must look at the previous sentences. We know from above that the NAACP is an activist group, so we can assume that the BPP is also an activist group.

12. A: Choice *A* is correct because the Black Panther Party is one entity; therefore, the possession should show the "Party's approach" with the apostrophe between the "y" and the "s." Choice *B* is incorrect because the word "Parties" should not be plural. Choice *C* is incorrect because the apostrophe indicates that the word "Partys" is plural. The plural of "party" is "parties." Choice *D* is incorrect because, again, the word "parties" should not be plural; instead, it is one unified party.

13. C: Choice *C* is correct because the passage is told in past tense, and "enabled" is a past tense verb. Choice *A*, "enable," is present tense. Choice *B*, "are enabling," is a present participle, which suggests a continuing action. Choice *D*, "will enable," is future tense.

14. D: Choice *D* is correct because the conjunction "and" is the best way to combine the two independent clauses. Choice *A* is incorrect because the word "he" becomes repetitive since the two clauses can be joined together. Choice *B* is incorrect because the conjunction "but" indicates a contrast, and there is no contrast between the two clauses. Choice *C* is incorrect because the introduction of the comma after "project" with no conjunction creates a comma splice.

15. C: The word "acheivement" is misspelled. Remember the rules for "*i* before *e* except after *c*." Choices *B* and *D*, "greatest" and "leader," are both spelled correctly.

16. B: Choice *B* is correct because it provides the correct verb tense and verb form. Choice *A* is incorrect; Hampton was not "held by a press conference"—rather, he held a press conference. The passage indicates that he "made the gangs agree to a nonaggression pact," implying that it was Hampton who was doing the speaking for this conference. Choice *C* is incorrect because, with this use of the sentence, it would create a fragment because the verb "holding" has no helping verb in front of it. Choice *D* is incorrect because it adds an infinitive ("to hold") where a past tense form of a verb should be.

17. A: Choice *A* is correct because it provides the most clarity. Choice *B* is incorrect because it doesn't name the group until the end, so the phrase "the group" is vague. Choice *C* is incorrect because it indicates that the BPP's popularity grew as a result of placing the group under constant surveillance, which is incorrect. Choice *D* is incorrect because there is a misplaced modifier; this sentence actually says that the FBI's influence and popularity grew, which is incorrect.

18. B: Choice *B* is correct. Choice *A* is incorrect because there should be an independent clause on either side of a semicolon, and the phrase "In 1976" is not an independent clause. Choice *C* is incorrect because there should be a comma after introductory phrases in general, such as "In 1976," and Choice *C* omits a comma. Choice *D* is incorrect because the sentence "In 1976." is a fragment.

19. C: Choice *C* is correct because the past tense verb "provided" fits in with the rest of the verb tense throughout the passage. Choice *A*, "will provide," is future tense. Choice *B*, "provides," is present tense. Choice *D*, "providing," is a present participle, which means the action is continuous.

20. D: The correct answer is Choice *D* because this statement provides the most clarity. Choice *A* is incorrect because the noun "Chicago City Council" acts as one, so the verb "are" should be singular, not plural. Choice *B* is incorrect because it is perhaps the most confusingly worded out of all the answer choices; the phrase "December 4" interrupts the sentence without any indication of purpose. Choice *C* is incorrect because it is too vague and leaves out *who* does the commemorating.

21. B: Choice *B* is correct. Here, a colon is used to introduce an explanation. Colons either introduce explanations or lists. Additionally, the quote ends with the punctuation inside the quotes, unlike Choice *C*.

22. A: The verb tense in this passage is predominantly in the present tense, so Choice *A* is the correct answer. Choice *B* is incorrect because the subject and verb do not agree. It should be "Education provides," not "Education provide." Choice *C* is incorrect because the passage is in present tense, and "Education will provide" is future tense. Choice *D* doesn't make sense when placed in the sentence.

23. D: The possessive form of the word "it" is "its." The contraction "it's" denotes "it is." Thus, Choice *A* is wrong. The word "raises" in Choice *B* makes the sentence grammatically incorrect. Choice *C* adds an apostrophe at the end of "its." While adding an apostrophe to most words would indicate possession, adding 's to the word "it" indicates a contraction.

24. C: The word *civilised* should be spelled *civilized.* The words "distinguishes" and "creatures" are both spelled correctly.

25. B: Choice *B* is correct because it provides clarity by describing what "myopic" means right after the word itself. Choice *A* is incorrect because the explanation of "myopic" comes before the word; thus, the meaning is skewed. It's possible that Choice *C* makes sense within context. However, it's not the *best* way to say this because the commas create too many unnecessary phrases. Choice *D* is confusingly worded. Using "myopic focus" is not detrimental to society; however, the way *D* is worded makes it seem that way.

26. C: Again, we see where the second paragraph can be divided into two parts due to separate topics. The paragraph's main focus is education addressing the mind, body, and soul. The first section, then, could end with the concluding sentence, "The human heart and psyche . . ." The next sentence to start a new paragraph would be "Education is a basic human right." The rest of this paragraph talks about what education is and some of its characteristics.

27. A: Choice *A* is correct because the phrase "others' ideas" is both plural and indicates possession. Choice *B* is incorrect because "other's" indicates only one "other" that's in possession of "ideas," which is incorrect. Choice *C* is incorrect because no possession is indicated. Choice *D* is incorrect because the word "other" does not end in *s*. Others's is not a correct form of the plural possessive word.

28. D: This sentence must have a comma before "although" because the word "although" is connecting two independent clauses. Thus, Choices *B* and *C* are incorrect. Choice *A* is incorrect because the second sentence in the underlined section is a fragment.

29. C: Choice *C* is the correct choice because the word "their" indicates possession, and the text is talking about "their students," or the students of someone. Choice *A*, "there," means at a certain place and is incorrect. Choice *B*, "they're," is a contraction and means "they are." Choice *D* is not a word.

30. B: Choice *B* uses all punctuation correctly in this sentence. In American English, single quotes should only be used if they are quotes within a quote, making choices *A* and *C* incorrect. Additionally, punctuation should go inside quotation marks with a few exceptions, making Choice D incorrect.

31. B: Choice *B* is correct because the conjunction "and" is used to connect phrases that are to be used jointly, such as teachers working hard to help students "identify salient information" and to "think critically." The conjunctions *so, but*, and *nor* are incorrect in the context of this sentence.

32. A: Choice *A* has consistent parallel structure with the verbs "read," "identify," and "determine." Choices *B* and *C* have faulty parallel structure with the words "determining" and "identifying." Choice *D* has incorrect subject/verb agreement. The sentence should read, "Students have to read . . . identify . . . and determine."

33. D: The correct choice for this sentence is that "they are . . . shaped by the influences." The prepositions "for," "to," and "with" do not make sense in this context. People are *shaped by*, not *shaped for, shaped to,* or *shaped with*.

34. A: To see which answer is correct, it might help to place the subject, "Teachers," near the verb. Choice *A* is correct: "Teachers . . . must strive" makes grammatical sense here. Choice B is incorrect because "Teachers . . . to strive" does not make grammatical sense. Choice C is incorrect because "Teachers must not only respect . . . but striving" eschews parallel structure. Choice *D* is incorrect because it is in past tense, and this passage is in present tense.

35. C: Choice *C* is correct because it uses an em-dash. Em-dashes are versatile. They can separate phrases that would otherwise be in parenthesis, or they can stand in for a colon. In this case, a colon would be another decent choice for this punctuation mark because the second sentence expands upon the first sentence. Choice *A* is incorrect because the statement is not a question. Choice *B* is incorrect because adding a comma here would create a comma splice. Choice *D* is incorrect because this creates a run-on sentence since the two sentences are independent clauses.

36. B: The best place for this sentence given all the answer choices is at the end of the first paragraph. Choice *A* is incorrect; the passage is told in chronological order, and leaving the sentence as-is defies that order, since we haven't been introduced to who raised George. Choice *C* is incorrect because this sentence is not an introductory sentence. It does not provide the main topic of the paragraph. Choice *D* is incorrect because again, it defies chronological order. By the end of paragraph two we have already gotten to George as an adult, so this sentence would not make sense here.

37. D: Out of these choices, a semicolon would be the best fit because there is an independent clause on either side of the semicolon, and the two sentences closely relate to each other. Choice *A* is incorrect because putting a comma between two independent clauses (i.e. complete sentences) creates a comma splice. Choice *B* is incorrect; omitting punctuation here creates a run-on sentence. Choice *C* is incorrect because an ellipses (. . .) is used to designate an omission in the text.

38. B: Choice *B* is the correct answer because it uses "ing" verbs as gerunds. Gerunds are "ing" words that stand in for nouns. The words "growing" and "depleting" are gerunds in this example. Choice *B* also uses the conjunction "and," whereas the other answer choices have comma splices.

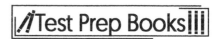

39. C: Choice *C* is correct because it keeps with the verb tense in the rest of the passage: past tense. Choice *A* is in present tense, which is incorrect. Choice *B* is present progressive, which means there is a continual action, which is also incorrect. Choice *D* is incorrect because "was returning" is past progressive tense, which means that something was happening continuously at some point in the past.

40. C: The correct choice is the subordinating conjunction, "When." We should look at the clues around the phrase to see what fits best. Carver left his money "when he died." Choice *A*, "Because," could perhaps be correct, but "When" is the more appropriate word to use here. Choice *B* is incorrect; "Although" denotes a contrast, and there is no contrast here. Choice *D* is incorrect because "Finally" indicates something at the very end of a list or series, and there is no series at this point in the text.

41. A: There should be no change here. Both underlined sentences are complete and do not need changing. Choice *B* is incorrect because there is no punctuation between the two independent clauses, it is considered a run-on. Choice *C* is incorrect because placing a comma between two independent clauses creates a comma splice. Choice *D* is incorrect. The underlined portion could *possibly* act with a colon. However, it's not the best choice, so omit Choice *D*.

42. C: Choice *C* is correct; the underlined phrase consists of part of an independent clause and a dependent clause ("Depending on which historian you seek.") The dependent clause cannot stand by itself. Thus, the best choice is to connect the two clauses with a comma. Choices *A* and *D* do not work because you must have two independent clauses on either side of a period as well as a semicolon. Choice *B* is incorrect because an exclamation point is used to show excitement and does not fit the tone here.

43. B: The most appropriate verb for this sentence is Choice *B*, "Leif consulted." Choice *A* is in present tense and therefore does not fit with the rest of the passage. Choice *C* is incorrect because "consulting" is present progressive tense and also does not fit with the consistent past tense of the passage. Choice *D*, "Leif was consulted with a merchant," doesn't make sense. Leif can consult with a merchant or be consulted by a merchant.

44. C: Choice *C* is correct. The subject and verb agree with each other (accounts describe), and there is no apostrophe because no possession is being shown. Choices *B* and *D* are incorrect because there is no possession—"accounts" is simply plural. Choice *A* is incorrect because the subject and verb do not agree with each other (accounts describes).

Math

1. B: To simplify this inequality, subtract 3 from both sides to get $-\frac{1}{2}x \geq -1$. Then, multiply both sides by -2 (remembering this flips the direction of the inequality) to get $x \leq 2$.

2. D: This problem involves a composition function, where one function is plugged into the other function. In this case, the $f(x)$ function is plugged into the $g(x)$ function for each x-value. The composition equation becomes:

$$g\big(f(x)\big) = 2^3 - 3(2^2) - 2(2) + 6$$

Simplifying the equation gives the answer:

$$g(f(x)) = 8 - 3(4) - 2(2) + 6$$

$$8 - 12 - 4 + 6 = -2$$

3. C: An *x*-intercept is the point where the graph crosses the *x*-axis. At this point, the value of *y* is 0. To determine if an equation has an *x*-intercept of -2, substitute -2 for *x,* and calculate the value of *y*. If the value of -2 for *x* corresponds with a *y*-value of 0, then the equation has an *x*-intercept of -2. The only answer choice that produces this result is:

$$\text{Choice } C \rightarrow 0 = (-2)^2 + 5(-2) + 6$$

4. C: The conditional frequency of a girl being potty-trained is calculated by dividing the number of potty-trained girls by the total number of girls:

$$34 \div 52 = 0.65$$

To determine the conditional probability, multiply the conditional frequency by 100:

$$0.65 \times 100 = 65\%$$

5. B: The volume of a cube is the length of the side cubed, and 3 inches cubed is 27 in³. Choice *A* is not the correct answer because that is 2×3 inches. Choice *C* is not the correct answer because that is 3×3 inches, and Choice *D* is not the correct answer because there was no operation performed.

6. B: The volume of a rectangular prism is the $length \times width \times height$, and $3cm \times 5cm \times 11cm$ is 165 cm³. Choice *A* is not the correct answer because that is $3cm + 5cm + 11cm$. Choice *C* is not the correct answer because that is 15^2. Choice *D* is not the correct answer because that is $3cm \times 5cm \times 10cm$.

7. A: The volume of a cylinder is $\pi r^2 h$, and $\pi \times 5^2 \times 10$ is $250\pi\ in^3$. Choice *B* is not the correct answer because that is $5^2 \times 2\pi$. Choice *C* is not the correct answer since that is $2 \times 5in \times 10\pi$. Choice *D* is not the correct answer because that is $10^2 \times 2in$.

8. D: This system of equations involves one quadratic function and one linear function, as seen from the degree of each equation. One way to solve this is through substitution. Solving for y in the second equation yields $y = x + 2$. Plugging this equation in for the y of the quadratic equation yields:

$$x^2 - 2x + x + 2 = 8$$

Simplifying the equation, it becomes:

$$x^2 - x + 2 = 8$$

Setting this equal to zero and factoring, it becomes:

$$x^2 - x - 6 = 0$$

$$(x - 3)(x + 2) = 0$$

Solving these two factors for x gives the zeros $x = 3, -2$. To find the y-value for the point, each number can be plugged in to either original equation. Solving each one for y yields the points $(3, 5)$ and $(-2, 0)$.

9. B: The slope will be given by $\frac{1-0}{2-0} = \frac{1}{2}$. The y-intercept will be 0, since it passes through the origin. Using slope-intercept form, the equation for this line is $y = \frac{1}{2}x$.

10. D: Area = length x width. The answer must be in square inches, so all values must be converted to inches. $\frac{1}{2}$ ft is equal to 6 inches. Therefore, the area of the rectangle is equal to:

$$6 \times \frac{11}{2} = \frac{66}{2} = 33 \text{ square inches}$$

11. B: The table shows values that are increasing exponentially. The differences between the inputs are the same, while the differences in the outputs are changing by a factor of 2. The values in the table can be modeled by the equation:

$$f(x) = 2^x$$

12. B: For the first card drawn, the probability of a King being pulled is $\frac{4}{52}$. Since this card isn't replaced, if a King is drawn first the probability of a King being drawn second is $\frac{3}{51}$. The probability of a King being drawn in both the first and second draw is the product of the two probabilities: $\frac{4}{52} \times \frac{3}{51} = \frac{12}{2652}$. This fraction, when divided by 12, equals $\frac{1}{221}$.

13. C: The picture demonstrates Angle-Side-Angle congruence. Choice *A* is not the correct answer because the picture does not show Side-Side-Side congruence. Choice *B* is not the correct answer because Angle-Angle-Angle does not even prove congruence. Choice *D* is not the correct answer because there is enough information to prove congruence.

14. D: The expression is three times the sum of twice a number and 1, which is $3(2x + 1)$. Then, 6 is subtracted from this expression.

15. C: The average is calculated by adding all six numbers, then dividing by 6. The first five numbers have a sum of 25. If the total divided by 6 is equal to 6, then the total itself must be 36. The sixth number must be $36 - 25 = 11$.

16. A: To expand a squared binomial, it's necessary to use the *First, Inner, Outer, Last Method*.

$$(2x - 4y)^2 = 2x \times 2x + 2x(-4y) + (-4y)(2x) + (-4y)(-4y)$$

$$4x^2 - 8xy - 8xy + 16y^2 = 4x^2 - 16xy + 16y^2$$

17. B: The zeros of this function can be found by using the quadratic formula:

$$x = \frac{-b \pm \sqrt{b^2 - 4ac}}{2a}$$

Identifying *a*, *b*, and *c* can be done from the equation as well because it is in standard form. The formula becomes:

$$x = \frac{0 \pm \sqrt{0^2 - 4(1)(4)}}{2(1)} = \frac{\sqrt{-16}}{2}$$

Since there is a negative underneath the radical, the answer is a complex number: $x = \pm 2i$.

18. D: The expression is simplified by collecting like terms. Terms with the same variable and exponent are like terms, and their coefficients can be added.

19. D: The slope is given by the change in y divided by the change in x. Specifically, it's:

$$slope = \frac{y_2 - y_1}{x_2 - x_1}$$

The first point is $(-5, -3)$, and the second point is $(0, -1)$. Work from left to right when identifying coordinates. Thus, the point on the left is point 1 $(-5, -3)$ and the point on the right is point 2 $(0, -1)$.

Now we need to just plug those numbers into the equation:

$$slope = \frac{-1 - (-3)}{0 - (-5)}$$

It can be simplified to:

$$slope = \frac{-1 + 3}{0 + 5}$$

$$slope = \frac{2}{5}$$

20. A: Finding the product means distributing one polynomial to the other so that each term in the first is multiplied by each term in the second. Then, like terms can be collected. Multiplying the factors yields the expression:

$$20x^3 + 4x^2 + 24x - 40x^2 - 8x - 48$$

Collecting like terms means adding the x^2 terms and adding the x terms. The final answer after simplifying the expression is:

$$20x^3 - 36x^2 + 16x - 48$$

21. B: The equation can be solved by factoring the numerator into $(x + 6)(x - 5)$. Since that same factor $(x - 5)$ exists on top and bottom, that factor cancels. This leaves the equation $x + 6 = 11$. Solving the equation gives the answer $x = 5$. When this value is plugged into the equation, it yields a zero in the denominator of the fraction. Since this is undefined, there is no solution.

22. A: The common denominator here will be $4x$. Rewrite these fractions as:

$$\frac{3}{x} + \frac{5u}{2x} - \frac{u}{4}$$

$$\frac{12}{4x} + \frac{10u}{4x} - \frac{ux}{4x}$$

$$\frac{12 + 10u - ux}{4x}$$

23. B: There are two zeros for the given function. They are $x = 0, -2$. The zeros can be found a number of ways, but this particular equation can be factored into:

$$f(x) = x(x^2 + 4x + 4)$$

$$x(x + 2)(x + 2)$$

By setting each factor equal to zero and solving for x, there are two solutions. On a graph, these zeros can be seen where the line crosses the x-axis.

24. A: The equation is *even* because:

$$f(-x) = f(x)$$

Plugging in a negative value will result in the same answer as when plugging in the positive of that same value.

The function:

$$f(-2) = \frac{1}{2}(-2)^4 + 2(-2)^2 - 6$$

$$8 + 8 - 6 = 10$$

yields the same value as:

$$f(2) = \frac{1}{2}(2)^4 + 2(2)^2 - 6$$

$$8 + 8 - 6 = 10$$

25. A: The three sides are shown to be congruent. Choice *B* is not the correct answer because Angle-Angle-Angle does not prove congruence. Choice *C* is not the correct answer because Angle-Side-Angle is not shown. Choice *D* is not the correct answer because there is enough information to prove the triangles are congruent, since all three sides are congruent (SSS).

26. C: Because x and y are complementary, the $\sin(x) = \cos(y)$. Choice *A* is not the correct answer because that is $1 - 0.8$. Choice *B* is not the correct answer because that is the negative value of $1 - 0.8$. Choice *D* is not the correct value because that is the positive value of the correct answer.

27. B: Because this isn't a right triangle, SOHCAHTOA can't be used. However, the law of cosines can be used. Therefore:

$$c^2 = a^2 + b^2 - 2ab \cos C = 19^2 + 26^2 - 2 \cdot 19 \cdot 26 \cdot \cos 42° = 302.773$$

Taking the square root and rounding to the nearest tenth results in $c = 17.4$.

28. C: Because the triangles are similar, the lengths of the corresponding sides are proportional. Therefore:

$$\frac{30 + x}{30} = \frac{22}{14} = \frac{y + 15}{y}$$

This results in the equation $14(30 + x) = 22 \cdot 30$ which, when solved, gives $x = 17.1$. The proportion also results in the equation $14(y + 15) = 22y$ which, when solved, gives $y = 26.3$.

29. B: The technique of completing the square must be used to change $4x^2 + 4y^2 - 16x - 24y + 51 = 0$ into the standard equation of a circle. First, the constant must be moved to the right-hand side of the equals sign, and each term must be divided by the coefficient of the x^2 term (which is 4). The x and y terms must be grouped together to obtain:

$$x^2 - 4x + y^2 - 6y = -\frac{51}{4}$$

Then, the process of completing the square must be completed for each variable. This gives:

$$(x^2 - 4x + 4) + (y^2 - 6y + 9) = -\frac{51}{4} + 4 + 9$$

The equation can be written as:

$$(x - 2)^2 + (y - 3)^2 = \frac{1}{4}$$

Therefore, the center of the circle is (2, 3) and the radius is:

$$\sqrt{\frac{1}{4}} = \frac{1}{2}$$

30. B: Given the area of the circle, the radius can be found using the formula $A = \pi r^2$. In this case, $49\pi = \pi r^2$, which yields $r = 7$ m. A central angle is equal to the degree measure of the arc it inscribes; therefore, $\angle x = 80°$. The area of a sector can be found using the formula:

$$A = \frac{\theta}{360°} \times \pi r^2$$

In this case:

$$A = \frac{80°}{360°} \times \pi(7)^2 = 10.9\pi \text{ m}$$

31. C: Plug in the values for *x* and *y* to discover that the solution works, which is:

$$(-3)^2 + (-4)^2 = 25$$

Choices *A* and *B* are not the correct answers since the solution works. Choice *D* is not the correct answer because there is enough information to tell where the given point lies on the circle.

32. A: This answer is correct because $100 - 64$ is 36, and taking the square root of 36 is 6. Choice *B* is not the correct answer because that is $10 + 8$. Choice *C* is not the correct answer because that is 8×10. Choice *D* is also not the correct answer because there is no reason to arrive at that number.

33. D: The addition rule is necessary to determine the probability because a 6 can be rolled on either roll of the die. The rule used is:

$$P(A \text{ or } B) = P(A) + P(B) - P(A \text{ and } B)$$

The probability of a 6 being individually rolled is $\frac{1}{6}$ and the probability of a 6 being rolled twice is:

$$\frac{1}{6} \times \frac{1}{6} = \frac{1}{36}$$

Therefore, the probability that a 6 is rolled at least once is:

$$\frac{1}{6} + \frac{1}{6} - \frac{1}{36} = \frac{11}{36}$$

34. C: $y = 40x + 300$. In this scenario, the variables are the number of sales and Karen's weekly pay. The weekly pay depends on the number of sales. Therefore, weekly pay is the dependent variable (y), and the number of sales is the independent variable (x). Each pair of values from the table can be written as an ordered pair (x, y): (2, 380), (7, 580), (4, 460), (8, 620). The ordered pairs can be substituted into the equations to see which creates true statements (both sides equal) for each pair. Even if one ordered pair produces equal values for a given equation, the other three ordered pairs must be checked. The only equation which is true for all four ordered pairs is $y = 40x + 300$:

$$380 = 40(2) + 300 \rightarrow 380 = 380$$

$$580 = 40(7) + 300 \rightarrow 580 = 580$$

$$460 = 40(4) + 300 \rightarrow 460 = 460$$

$$620 = 40(8) + 300 \rightarrow 620 = 620$$

35. C: The area of the shaded region is the area of the square, minus the area of the circle. The area of the circle will be πr^2. The side of the square will be $2r$, so the area of the square will be $4r^2$. Therefore, the difference is:

$$4r^2 - \pi r^2 = (4 - \pi)r^2$$

36. B: The car is traveling at a speed of five meters per second. On the interval from one to three seconds, the position changes by ten meters. By making this change in position over time into a rate, the speed becomes ten meters in two seconds or five meters in one second.

37. B: For an ordered pair to be a solution to a system of inequalities, it must make a true statement for BOTH inequalities when substituting its values for x and y. Substituting $(-3, -2)$ into the inequalities produces $(-2) > 2(-3) - 3 \rightarrow -2 > -9$ and $(-2) < -4(-3) + 8 \rightarrow -2 < 20$. Both are true statements.

38. D: The shape of the scatterplot is a parabola (U-shaped). This eliminates Choices A (a linear equation that produces a straight line) and C (an exponential equation that produces a smooth curve upward or downward). The value of a for a quadratic function in standard form ($y = ax^2 + bx + c$) indicates whether the parabola opens up (U-shaped) or opens down (upside-down U). A negative value for a produces a parabola that opens down; therefore, Choice B can also be eliminated.

39. C: The cosine of 45° is equal to .7071. Choice *A* is not the correct answer because the cosine of 15° is .9659. Choice *B* is not the correct answer because the cosine of 30° is .8660. Choice *D* is not correct because the cosine of 90° is 0.

40. B: The tangent of 30° is 1 over the square root of 3. Choice *A* is not the correct answer because the tangent of 15° is .2679. Choice *C* is not the correct answer because the tangent of 45° is 1. Choice *D* is not the correct answer because the tangent of 90° is undefined.

41. D: Exponential functions can be written in the form: $y = a \cdot b^x$. The equation for an exponential function can be written given the *y*-intercept (*a*) and the growth rate (*b*). The *y*-intercept is the output (*y*) when the input (*x*) equals zero. It can be thought of as an "original value," or starting point. The value of *b* is the rate at which the original value increases ($b > 1$) or decreases ($b < 1$). In this scenario, the *y*-intercept, *a*, would be $1200, and the growth rate, *b*, would be 1.01 (100% of the original value + 1% interest = 101% = 1.01).

42. C: $x = 150$

Set up the initial equation.

$$\frac{2x}{5} - 1 = 59$$

Add 1 to both sides.

$$\frac{2x}{5} - 1 + 1 = 59 + 1$$

Multiply both sides by $\frac{5}{2}$.

$$\frac{2x}{5} \times \frac{5}{2} = 60 \times \frac{5}{2} = 150$$

$$x = 150$$

43. B: To simplify the given equation, the first step is to make all exponents positive by moving them to the opposite place in the fraction. This expression becomes:

$$\frac{4b^3b^2}{a^1a^4} \times \frac{3a}{b}$$

Then the rules for exponents can be used to simplify. Multiplying the same bases means the exponents can be added. Dividing the same bases means the exponents are subtracted.

44. D: Dividing rational expressions follows the same rule as dividing fractions. The division is changed to multiplication, and the reciprocal is found in the second fraction. This turns the expression into:

$$\frac{5x^3}{3x^2y} \times \frac{3y^9}{25}$$

Multiplying across and simplifying, the final expression is:

$$\frac{xy^8}{5}$$

45. D: Using SOHCAHTOA, tangent is $\frac{y}{x}$ for the special triangles. Since the value needs to be negative one, the angle must be some form of 45 degrees or $\frac{\pi}{4}$. The value is negative in the second and fourth quadrant, so the answer is $\frac{3\pi}{4}$ and $\frac{7\pi}{4}$.

46. D: Three girls for every two boys can be expressed as a ratio: 3:2. This can be visualized as splitting the school into 5 groups: 3 girl groups and 2 boy groups. The number of students which are in each group can be found by dividing the total number of students by 5:

650 divided by 5 equals 1 part, or 130 students per group

To find the total number of girls, multiply the number of students per group (130) by the number of girl groups in the school (3). This equals 390, which is answer D.

47. A: The method to equate two chord lengths is to multiply the two segments of each chord by each other and then set the resulting products equal to each other to find the value of x:

$$7(2x + 2) = 6(x + 5)$$

$$14x + 14 = 6x + 30$$

$$16 = 8x \rightarrow x = 2$$

Then, substitute the value of x into one of the chord segment equations and add them together to find the length of the chords:

$$7 + (2x + 2) \rightarrow 7 + (2(2) + 2) = 13$$

Choice B is not the correct answer because 2 is the solution for x, not the length of the chord. Choice C is not the correct answer because 42 is the chord length if the two segments are multiplied together rather than added. Choice D is not the correct answer because 6 is the value of only one segment not the total chord length.

48. C: One apple/orange pair costs $3 total. Therefore, Jan bought $\frac{90}{3} = 30$ total pairs, and hence, she bought 30 oranges.

49. D: First, the train's journey in the real word is $3 \times 50 = 150$ miles. On the map, 1 inch corresponds to 10 miles, so there is $\frac{150}{10} = 15$ inches on the map.

50. B: The total trip time is $1 + 3.5 + 0.5 = 5$ hours. The total time driving is $1 + 0.5 = 1.5$ hours. So, the fraction of time spent driving is $\frac{1.5}{5}$ or $\frac{3}{10}$. To get the percentage, convert this to a fraction out of 100. The numerator and denominator are multiplied by 10, with a result of $\frac{30}{100}$. The percentage is the numerator in a fraction out of 100, so 30%.

51. B: The formula for the volume of a cylinder is $\pi r^2 h$, where r is the radius and h is the height. The diameter is twice the radius, so these barrels have a radius of 1 foot. That means each barrel has a volume of:

$$\pi \times 1^2 \times 3 = 3\pi \text{ cubic feet}$$

Since there are three of them, the total is:

$$3 \times 3\pi = 9\pi \text{ cubic feet}$$

52. C: 216cm. Because area is a two-dimensional measurement, the dimensions are multiplied by a scale that is squared to determine the scale of the corresponding areas. The dimensions of the rectangle are multiplied by a scale of 3. Therefore, the area is multiplied by a scale of 3^2 (which is equal to 9):

$$24cm \times 9 = 216cm$$

53.

First solve for x, y, and z. So, $3x = 24$, $x = 8$, $6y = 24$, $y = 4$, and $-2z = 24$, $z = -12$. This means the equation would be $4(8)(4) + (-12)$, which equals 116.

54.

Add 3 to both sides to get $4x = 8$. Then divide both sides by 4 to get $x = 2$.

55.

To solve this correctly, keep in mind the order of operations with the mnemonic PEMDAS (Please Excuse My Dear Aunt Sally). This stands for Parentheses, Exponents, Multiplication, Division, Addition, Subtraction. Taking it step by step, solve the parentheses first:

$$4 \times 7 + (4)^2 \div 2$$

Then, apply the exponent:

$$4 \times 7 + 16 \div 2$$

Multiplication and division are both performed next:

$$28 + 8 = 36$$

56.

			1	5
⊖	⊖	⊖	⊖	⊖
	⊙	⊙	⊙	⊙
	⓪	⓪	⓪	⓪
	①	①	●	①
	②	②	②	②
	③	③	③	③
	④	④	④	④
	⑤	⑤	⑤	●
	⑥	⑥	⑥	⑥
	⑦	⑦	⑦	⑦
	⑧	⑧	⑧	⑧
	⑨	⑨	⑨	⑨

Follow the *order of operations* in order to solve this problem. Solve the parentheses first, and then follow the remainder as usual.

$$(6 \times 4) - 9$$

This equals $24 - 9$ or 15.

57.

	8	7	.	5
⊖	⊖	⊖	⊖	⊖
	⊙	⊙	●	⊙
	⓪	⓪	⓪	⓪
	①	①	①	①
	②	②	②	②
	③	③	③	③
	④	④	④	④
	⑤	⑤	⑤	●
	⑥	⑥	⑥	⑥
	⑦	●	⑦	⑦
	●	⑧	⑧	⑧
	⑨	⑨	⑨	⑨

For an even number of total values, the *median* is calculated by finding the *mean* or average of the two middle values once all values have been arranged in ascending order from least to greatest. In this case, $(92 + 83) \div 2$ would equal the median 87.5.

58.

Start with the original equation: $x^2 - 2xy + 2y^2$, then replace each instance of x with a 2, and each instance of y with a 3 to get:

$$2^2 - 2 \times 2 \times 3 + 2 \times 3^2$$

$$4 - 12 + 18 = 10$$

Dear PSAT Test Taker,

We would like to start by thanking you for purchasing this study guide for your PSAT exam. We hope that we exceeded your expectations.

Our goal in creating this study guide was to cover all of the topics that you will see on the test. We also strove to make our practice questions as similar as possible to what you will encounter on test day. With that being said, if you found something that you feel was not up to your standards, please send us an email and let us know.

We would also like to let you know about other books in our catalog that may interest you.

SAT

This can be found on Amazon: amazon.com/dp/1628458984

ACT

amazon.com/dp/1628458844

ACCUPLACER

amazon.com/dp/162845945X

CLEP College Composition

amazon.com/dp/1628454199

We have study guides in a wide variety of fields. If the one you are looking for isn't listed above, then try searching for it on Amazon or send us an email.

Thanks Again and Happy Testing!
Product Development Team
info@studyguideteam.com

FREE Test Taking Tips DVD Offer

To help us better serve you, we have developed a Test Taking Tips DVD that we would like to give you for FREE. **This DVD covers world-class test taking tips that you can use to be even more successful when you are taking your test.**

All that we ask is that you email us your feedback about your study guide. Please let us know what you thought about it – whether that is good, bad or indifferent.

To get your **FREE Test Taking Tips DVD**, email freedvd@studyguideteam.com with "FREE DVD" in the subject line and the following information in the body of the email:

    a. The title of your study guide.

    b. Your product rating on a scale of 1-5, with 5 being the highest rating.

    c. Your feedback about the study guide. What did you think of it?

    d. Your full name and shipping address to send your free DVD.

If you have any questions or concerns, please don't hesitate to contact us at freedvd@studyguideteam.com.

Thanks again!

Made in the USA
Las Vegas, NV
26 August 2021

29038974R00149